Antenna Analysis and Design Using FEKO Electromagnetic Simulation Software

Antenna Analysis and Design Using FEKO Electromagnetic Simulation Software

Atef Z. Elsherbeni

Electrical Engineering and Computer Science Department,
Colorado School of Mines

Payam Nayeri

Electrical Engineering and Computer Science Department,
Colorado School of Mines

C. J. Reddy

EM Software & Systems (USA), Inc.

SciTECH
PUBLISHING
an imprint of the IET

Edison, NJ
theiet.org

Published by SciTech Publishing, an imprint of the IET.
www.scitechpub.com
www.theiet.org

Editor: Dudley R. Kay

10 9 8 7 6 5 4 3 2 1

ISBN 978-1-61353-205-8 (hardback)

Typeset in India by MPS Limited

Contents

Preface

Objective

The objective of this book is to introduce students and interested researchers to antenna design and analysis using the popular commercial electromagnetic software FEKO. Because this book is tutorial in nature, it is primarily intended for students working in the field of antenna analysis and design. However, its wealth of hands-on design examples as well as simulation details also makes the book a valuable reference for practicing engineers. Readers must be familiar with the basics of antenna theory, but electrical engineering students taking an introductory course in antenna engineering can also benefit from this book as a supplementary text.

Use of FEKO

The simulation files in this book are generated using the commercial electromagnetic software FEKO due to its numerous advantages. In particular, this software provides multiple solvers that can be chosen for a fast and efficient analysis depending on the type of antenna. Moreover, FEKO is providing free classroom licenses to universities, which can be used to simulate most of the antenna configurations studied in this book. This book illustrates how to set up the simulation model in FEKO for a variety of antenna types and configurations. FEKO version 6.2 is used for the development of the FEKO project files in this edition of the book.

The Book's Key Strengths

The first three chapters of the book present the basics of antenna simulation in a detailed, understandable, and easy-to-follow procedure through the study of the simplest types of radiators: dipole and loop antennas. Theoretical analysis and full-wave simulation results using FEKO are also compared together to give readers a better understanding of the theory, limitations, and approximations that are usually made in the design of antennas based on theoretical analysis. This will also build readers' fundamental knowledge to effectively use antenna simulation software. A variety of antenna configurations are then studied in Chapters 4 to 11, and, where appropriate, discussions or comparison with the theoretical solutions are also presented.

In addition, of paramount importance is the book's visualization of the antenna current distribution, radiation patterns, and other radiation characteristics. A proper analysis of the radiation characteristics through these visualizations serves as a powerful educational tool to fully understand the radiation behavior of antennas.

At the end of each chapter, a set of exercises related to the introduced antenna types will help readers assess their understanding of the covered material.

Errors and Suggestions

The authors welcome any feedback from the readers to possible errors and to suggestions for improving the presentation of topics and examples in this book. If needed, errata will be posted on the publisher's website.

Reviewer Acknowledgment

The authors gratefully acknowledge the time and expertise of the technical reviewers who assisted in enhancing the development of this book. The thoughtful comments and suggestions provided by these academic and industry experts greatly helped the authors to improve the quality and usefulness of the material presented in this book.

Dr. Randy Jost, *Ball Aerospace & Technologies Corp.*
Prof. Randy Haupt, *Colorado School of Mines*
Dr. William O'Keefe Coburn, *US Army Research Laboratory*
Dr. Alan J. Fenn, *Lincoln Laboratory, Massachusetts Institute of Technology*

Author Biographies

Atef Z. Elsherbeni graduated with honors from Cairo University, Egypt, with a B.Sc. in electronics and communications (1976), a B.Sc. in applied physics (1979), and an M.Eng. in electrical engineering (1982). He earned his Ph.D. in electrical engineering from Manitoba University, Winnipeg, Manitoba, Canada, in 1987. From March 1980 to December 1982, Dr. Elsherbeni was a part-time software and system design engineer at the Automated Data System Center in Cairo, Egypt. From January to August 1987, he was a Post Doctoral Fellow at Manitoba University. He joined the faculty at the University of Mississippi in August 1987 as assistant professor of electrical engineering and advanced to associate professor in 1991 and then to professor in 1997. He was director of the School of Engineering CAD Lab from 2002 to 2013, director of the Center for Applied Electromagnetic Systems Research (CAESR) from 2011 to 2013, and associate dean of engineering for Research and Graduate Programs from 2009 to 2013. He became the Dobelman Distinguished Chair and professor of electrical engineering and computer science at Colorado School of Mines in August 2013. He was appointed adjunct professor in the Department of Electrical Engineering and Computer Science of the L. C. Smith College of Engineering and Computer Science at Syracuse University in 2004. He spent a sabbatical term in 1996 at the Electrical Engineering Department, University of California at Los Angeles (UCLA) and was visiting professor at Magdeburg University during summer 2005 and at Tampere University of Technology in Finland in summer 2007. In 2009 he was selected as Finland Distinguished Professor by the Academy of Finland and TEKES.

Dr. Elsherbeni has received the 2013 Applied Computational Electromagnetics Society (ACES) Technical Achievements Award, the 2012 University of Mississippi Distinguished Research and Creative Achievement Award, the 2006 and the 2011 School of Engineering Senior Faculty Research Award for Outstanding Performance in Research, and the 2005 School of Engineering Faculty Service Award for Outstanding Performance in Service. Other accolades include the 2004 ACES Valued Service Award for Outstanding Service as 2003 ACES Symposium Chair, the Mississippi Academy of Science 2003 Outstanding Contribution to Science Award, the 2002 IEEE Region 3 Outstanding Engineering Educator Award, the 2002 School of Engineering Outstanding Engineering Faculty Member of the Year Award, the 2001 ACES Exemplary Service Award for leadership and contributions as electronic publishing managing editor 1999–2001, the 2001 Researcher/Scholar of the Year award in the Department of Electrical Engineering at the University of

Mississippi, and the 1996 Outstanding Engineering Educator of the IEEE Memphis Section.

Over the last 26 years, Dr. Elsherbeni participated in acquiring over $10 million to support his research dealing with electromagnetic wave scattering and diffraction of dielectric and metal objects; finite difference time domain analysis of antennas and microwave devices; field visualization and software development for electromagnetic education; interactions between electromagnetic waves and the human body, radio frequency identification (RFID) and sensor Integrated FRID systems, reflector and printed antennas and antenna arrays for radars; unmanned aerial vehicles (UAVs); and personal communication systems, antennas for wideband applications, and measurements of antenna characteristics and material properties. He has coauthored 148 technical journal articles and 28 book chapters; has contributed to over 350 professional presentations; and has offered 26 short courses and 30 invited seminars. He is coauthor of *The Finite Difference Time Domain Method for Electromagnetics with Matlab Simulations* (Scitech, 2009), *Antenna Design and Visualization Using MATLAB* (Scitech, 2006), *MATLAB Simulations for Radar Systems Design* (CRC Press, 2003), *Electromagnetic Scattering Using the Iterative Multiregion Technique* (Morgan & Claypool, 2007), *Electromagnetics and Antenna Optimization Using Taguchi's Method* (Morgan & Claypool, 2007), *Scattering Analysis of Periodic Structures Using Finite-Difference Time-Domain Method* (Morgan & Claypool, 2012), and *Multiresolution Frequency Domain Technique for Electromagnetics* (Morgan & Claypool, 2012) and is the primary author of the chapters "Handheld Antennas" and "The Finite Difference Time Domain Technique for Microstrip Antennas" in *Handbook of Antennas in Wireless Communications* (CRC Press, 2001). He has served as advisor or co-advisor for 33 M.S. and 14 Ph.D. students.

Dr. Elsherbeni is a fellow member of the Institute of Electrical and Electronics Engineers (IEEE) and the Applied Computational Electromagnetic Society (ACES). He is editor-in-chief for *ACES Journal* and past associate editor for *Radio Science Journal*. He serves on the editorial board of the book series *Progress in Electromagnetic Research* and *Electromagnetic Waves and Applications Journal*. He has served as chair of the Engineering and Physics Division of the Mississippi Academy of Science and the Educational Activity Committee for the IEEE Region 3 Section.

Payam Nayeri received his B.S. in applied physics from Shahid Beheshti University, Tehran, in 2004, his M.S. in electrical engineering from Iran University of Science and Technology, Tehran, in 2007, and his Ph.D. in electrical engineering from the University of Mississippi, in 2012. From 2008 to 2012 he was a graduate student researcher in the Center for Applied Electromagnetic Systems Research (CAESR) at the University of Mississippi. From 2012 to 2013 he was postdoctoral research associate and instructor in the Department of Electrical Engineering there. In January 2014 he became a postdoctoral fellow in the Department of Electrical Engineering and Computer Science at the Colorado School of Mines. His research interests include array and reflectarray antennas, reflector and lens antennas, multibeam and beam-scanning arrays, microstrip antennas, computational methods, and optimization in electromagnetics, and antenna measurement techniques.

Dr. Nayeri is a member of IEEE, Sigma Xi, Phi Kappa Phi, and the Applied Computational Electromagnetic Society (ACES). He has been the recipient of several prestigious awards, including the IEEE Antennas and Propagation Society Doctoral Research Award in 2010, the University of Mississippi Graduate Achievement Award in Electrical Engineering in 2011, and ACES's Best Student Paper Award of the 29th International Review of Progress in Applied Computational Electromagnetics in 2013. He currently serves as a technical reviewer for several journal and conference proceedings including *IEEE Transactions on Antennas and Propagation*, *IEEE Antennas and Wireless Propagation Letters*, *IET Microwaves, Antennas and Propagation Journal*, *Applied Computational Electromagnetics Society Journal*, *European Conference on Antennas and Propagation* (EUCAP), and *IEEE International Symposium on Antennas and Propagation* (AP-S).

C. J. Reddy received his B.Tech. in electronics and communications engineering from Regional Engineering College (now National Institute of Technology), Warangal, India, in 1983. He received his M.Tech. (1986) in microwave and optical communication engineering and his Ph.D. (1988) in electrical engineering from Indian Institute of Technology, Kharagpur. From 1987 to 1991 he worked as scientific officer at SAMEER (India) and participated in radar system design and development. In 1991, he was awarded a visiting fellowship from NSERC to conduct research at the Communications Research Center in Ottawa, Canada. In 1993 he was awarded a US National Research Council research associateship to conduct research in computational electromagnetics at NASA's Langley Research Center, Hampton, Virginia, during which time he was research professor at Hampton University from 1995 to 2000. During this time, he developed various finite element method codes for electromagnetics. He also worked on antenna design and simulation for automobiles and aircraft structures. Development of his hybrid finite element method/method of moments/geometrical theory of diffraction code for cavity-backed aperture antenna analysis received a Certificate of Recognition from NASA.

Since 2000, Dr. Reddy has been president and chief technical officer of Applied EM Inc. (http://www.appliedem.com), a small company specializing in the design and development of innovative antenna solutions and computational electromagnetics. There he has successfully led many small business innovative research projects for the United States Department of Defense (DoD). Some of the technologies developed under these projects are being considered for transition to the DoD. Dr. Reddy also serves as president of

EM Software & Systems (USA) Inc. (http://www.emssusa.com), where he is leading the marketing and support of commercial 3D electromagnetic simulation software, FEKO (http://www.feko.info), in North America.

Dr. Reddy is senior member of the Institute of Electrical and Electronics Engineers (IEEE) and Antenna Measurement Techniques Association (AMTA). He is fellow of Applied Computational Electromagnetic Society (ACES) and served as a member of its board of directors from 2006 to 2012. He has published 31 journal papers, 54 conference papers, and 17 NASA technical reports. Dr. Reddy was general chair of the ACES 2011 conference held in Williamsburg, Virginia, March 27–31, 2011.

Introduction to Antennas

1.1 | THE BASICS OF ANTENNAS

For centuries humans have been devising various methods to meet their need for long-distance communication. Early attempts relied on voice (such as drum and horn) and visual signals (such as flag and smoke), the latter of which uses the visible part of the electromagnetic spectrum. In the early nineteenth century, however, the discovery of electromagnetic waves began a whole new era for communicating over long distances.

The two components of a basic communication system are the transmitter and receiver. For radio frequency (RF) communications, these two components may be connected directly (with a wire) or indirectly (for wireless communication). For the latter, the transmitter (receiver) must have the ability to radiate (receive) radio waves. These RF devices are known as antennas. Antennas are used to transport electromagnetic energy from the transmitting source to the antenna or from the antenna to the receiver. Heinrich Hertz performed the first experimental demonstration of wireless communication in 1887. He designed a dipole antenna as the transmitter and received the signal with a loop antenna. In 1901, Guglielmo Marconi was the first to communicate wirelessly over a long distance via an RF signal across the Atlantic (Figure 1-1).

These days, different types of antennas for a wide range of applications have become an integral and important part of almost any electronic communication device. While in general all antennas are devices for receiving and transmitting

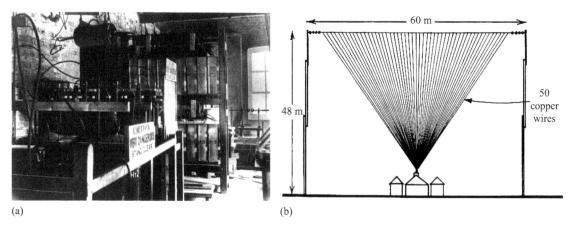

FIGURE 1-1 ▪ The first transatlantic radio communication [1]: (a) The Marconi Company transmitter at Poldhu, Cornwall, circa 1901. (b) The temporary fan-shaped antenna erected at Poldhu for the transatlantic experiment.

electromagnetic power, different applications require different functionalities. One way to categorize antenna types is by their gain which is a measure of the ratio of the radiation intensity in a given direction to the radiation intensity that would be obtained if the power accepted by the antenna were radiated isotropically. For applications where the two antennas are in proximity, such as wireless local area networks or cellular communication, antennas with low to moderate gain are usually used, such as dipole and patch. For very long-distance communication such as satellite or space communications, antennas with high gain are required, typically reflector and array. In this case a direct line of sight between the receiving and transmitting antenna is necessary to establish the communication link.

A few photographs of some commercial antennas are given in Figure 1-2. In addition to the gain requirements, the operating frequency of the antenna is determined by certain regulations. The Federal Communications Commission (FCC) designates certain frequency bands for commercial communications and also military applications. As such, another task for the antenna engineer is to design the antenna so that it can cover the required operating band. Many antennas such as dipoles and patches usually have a narrow bandwidth; hence, a challenge for the antenna engineer is to improve the bandwidth of these designs by using broadband techniques to satisfy the design requirements.

In addition to these design considerations for antennas, in practical applications surrounding objects have a notable effect on design performance. For example, a cellular phone's antenna must be placed in a package surrounded by several other electronic components as well as metallic objects. Thus, antenna engineers must design it so it can operate in the real environment.

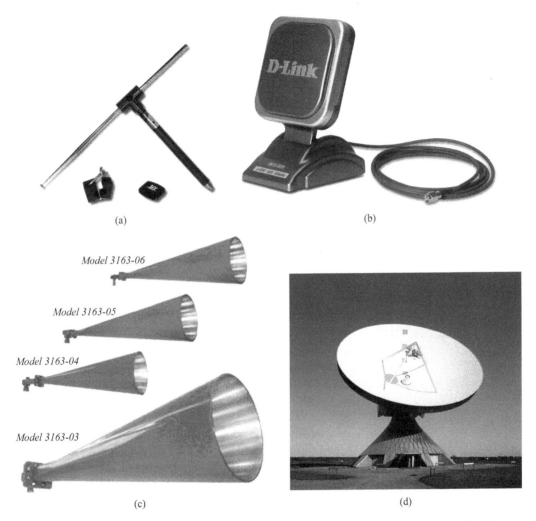

FIGURE 1-2 ■ (a) Dipole antenna [2]. (b) Patch antenna [3]. (c) Conical horn antennas [4]. (d) A high-gain parabolic antenna for satellite communication in Raisting, Germany [5].

Another frontier in antenna research is antenna configuration in communication networks. In the simplest form of a wireless link, that is, a single-input, single-output (SISO) system, one antenna is used for transmitting (receiving) the signal and another one for receiving (transmitting). Wireless communication, however, has now evolved so that multiple antennas forming a wireless network are used for transmitting and receiving. These systems are known as multiple-input, multiple-output (MIMO), and their antennas have certain requirements for isolation (Figure 1-3).

FIGURE 1-3 ▪
A MIMO antenna
system [6].

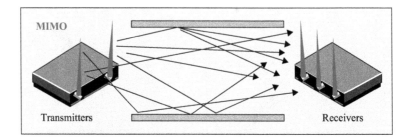

In this book we outline some specific challenges in antenna design that are not addressed in traditional antenna textbooks. The key to addressing them lies in properly understanding the antenna's radiation mechanism. In other words, how do antennas radiate, and what affects their radiation performance?

From physics we know that an accelerated charge radiates an electromagnetic wave. This is because an accelerated charge creates a disturbance in the electromagnetic fields that propagates away from the source. While in practice a source that generates accelerated charges may not be realized, a time-varying current source (derivate of charge with respect to time) that can generate an RF signal is available and can be used to generate this condition. The time-varying currents on the antenna produce electromagnetic radiation, which is governed by Maxwell's equations. In other words, the radiation pattern of the antenna is a manifest of its current. Consequently, controlling the shape of the radiation pattern is achieved using electromagnetic devices (antennas) that control the time-varying current. The previous brief discussion illustrates the necessity of studying and understanding the time-varying current behavior of the antenna, and throughout this book we will focus on this with the aim of providing a better understanding of the radiation mechanism of various types of antennas.

▮▮ 1.2 | ABOUT THIS BOOK

In the field of antenna engineering, theoretical analysis is of paramount importance in understanding the basics of the radiation mechanism. This is the focus of the majority of the books in the field (e.g., [7,8]) and will provide readers with a deeper knowledge of necessary concepts. The focus of this book, though, is on antenna simulation procedure and associated challenges to yield a good design.

It is significantly important in antenna design to find an accurate solution to the radiation problem. While the basic concept of antennas for transmitting and receiving an RF signal is well known, closed form exact analytical solutions to many antenna problems are not possible. Following the theoretical analysis methods, several approximations are typically made to simplify and solve the problem, but this consequently limits the accuracy of the solution. Almost exact solutions to antenna problems can be obtained numerically, however, which necessitates the use of advanced numerical electromagnetic methods and efficient computational software. In recent years, advances in computer hardware technology have notably improved computational power. This trend has always been positive but has now matured enough to a level that high-performance computational resources are available widely. Consequently, many computationally challenging antenna design problems that seemed impossible to unravel a few years ago can now be solved on personal computers.

In general, for a practical antenna design theoretical solutions are considered to be a starting point. To achieve a desirable and practical performance, though, antenna dimensions and other properties must be fine-tuned. Here we aim at providing a book that effectively combines the basics of the theoretical solutions and gives readers real hands-on design examples. Various antenna types are studied in this book, starting from simple configurations such as dipoles and loops and eventually leading to more complicated and practical designs such as broadband microstrip patches and high-gain reflector antennas. Analysis will be conducted using the popular electromagnetic simulator FEKO [9], and analytical expressions will be given whenever possible.

Chapter 2 and Chapter 3 begin with the most basic types of antennas: wire dipoles and loop antennas. Since these chapters serve as an introduction, more emphasis will be given on the basic details in the simulations to familiarize readers with the fundamental concepts and requirements of full-wave simulations. Later chapters will cover most of the commonly considered antenna types such as patches, horns, and reflectors. In Chapter 5 we will review the basics of microstrip based feed networks and then examine several examples of impedance transformers, power dividers, and couplers. Chapter 11 covers array antennas and presents several examples of dipole and patch type arrays.

In each chapter, the related electromagnetic concepts and fundamental antenna equations are first reviewed, and if applicable, comparisons between the analytical solution and full-wave simulations are presented and discussed. The effective utilization of full-wave electromagnetic simulations for every antenna configuration that is studied provides readers not only with the guidelines for antenna design and simulations but also with an illustrative visualization of antenna radiation patterns, radiating currents, and other radiation characteristics that will be very beneficial for both educational purposes and practical applications.

Wire Dipole and Monopole Antennas

2.1 | INTRODUCTION

Wire dipole antennas are the oldest, simplest, and also one of the cheapest antenna configurations. In addition, the half-wavelength dipole antenna is perhaps also the most widely used for low-gain applications. Detailed studies on the radiation characteristics of dipole antennas are available in the literature [7,8]. The ease with which they are designed and analyzed makes them a suitable starting point for students interested in the field of antenna engineering. In this chapter, we briefly study some of the fundamentals of dipole antennas, show a detailed procedure for designing dipole antennas using the commercial software package FEKO [9], and give several examples.

▮▮▮ 2.2 | INFINITESIMAL, SMALL, AND FINITE LENGTH DIPOLE ANTENNAS

Figure 2-1 shows a linear wire positioned symmetrically at the origin of the coordinate system and normal with respect to the x-y plane (i.e., oriented along the z axis).

FIGURE 2-1 ▪
Geometrical
arrangement of a
wire dipole antenna
and an observation
point, P.

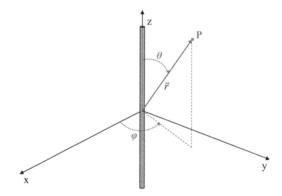

Here we assume that the wire is made of a perfect electric conductor material. Since the wire carries only electric currents, only the vector potential function, **A**, needs to be evaluated to compute the far-field radiation pattern. This is given by [7]

$$\mathbf{A}(x,y,z) = \frac{\mu}{4\pi} \int_c \mathbf{I_e}(x',y',z') \frac{e^{-jkR}}{R} dl', \qquad (2\text{-}1)$$

where (x, y, z) are the observation point coordinates, $\mathbf{I_e}$ is the current vector, (x', y', z') are the coordinates of the current distribution on the wire (source), R is the distance from any point on the source to the observation point, and k is the wavenumber. Since the current is only along the z direction, this equation can be further simplified, where in general this depends on the wire (dipole) length, l. In this chapter, the dipole wire radius is assumed to be much smaller than the wavelength. Dipole antennas with thick wires, also known as cylindrical dipoles, will be studied in a subsequent section.

For an infinitesimal wire element, $l \ll \lambda$, where λ is the wavelength of the generated radiation field, the current along the wire is assumed to be constant (i.e., I_0). In addition, the distance from a point on the wire to the observation point can be approximated for far field as

$$R = \sqrt{(x - x')^2 + (y - y')^2 + (z - z')^2} \approx \sqrt{x^2 + y^2 + z^2} = r. \qquad (2\text{-}2)$$

With these approximations, the vector potential function in (2-1) can be written as

$$\mathbf{A}(x,y,z) = \hat{z}\frac{\mu I_0 l}{4\pi r}e^{-jkr}. \tag{2-3}$$

Usually a constant current distribution on a wire element is not considered unless the wire length is less than $\lambda/50$. A more accurate and common current distribution is a triangular shape current on the wire with typical length in the range of $\lambda/50$ to $\lambda/10$. Mathematically this is given by

$$\mathbf{I}_e(x'=0,y'=0,z') = \begin{cases} \hat{z}I_0\left(1-\dfrac{2}{l}z'\right), & 0 \leq z' \leq l/2 \\[2mm] \hat{z}I_0\left(1+\dfrac{2}{l}z'\right), & -l/2 \leq z' \leq 0 \end{cases}. \tag{2-4}$$

The corresponding vector potential can then be simplified to

$$\mathbf{A}(x,y,z) = \frac{\mu}{4\pi}\left[\hat{z}\int_{-l/2}^{0}I_0\left(1+\frac{2}{l}z'\right)\frac{e^{-jkR}}{R}dz' + \hat{z}\int_{0}^{l/2}I_0\left(1-\frac{2}{l}z'\right)\frac{e^{-jkR}}{R}dz'\right]. \tag{2-5}$$

Similarly, in this case R can be approximated by $R \approx r$, since the wire length is relatively very small.

For wire dipole antennas with finite length, usually larger than $\lambda/10$, a good approximation of the current distribution is given by a sinusoidal distribution. The mathematical model of the sinusoidal current distribution is given by

$$\mathbf{I}_e(x'=0,y'=0,z') = \begin{cases} \hat{z}\,I_0\,\sin\left[k\left(\dfrac{l}{2}-z'\right)\right], & 0 \leq z' \leq l/2 \\[2mm] \hat{z}\,I_0\,\sin\left[k\left(\dfrac{l}{2}+z'\right)\right], & -l/2 \leq z' \leq 0 \end{cases}. \tag{2-6}$$

It is obvious that with this current distribution it becomes quite complicated to compute the vector potential function. To analyze the radiation pattern here, the finite length dipole is subdivided into a number of infinitesimal dipoles, with the appropriate constant current distribution on each segment. More details on the analytical procedure for computing the radiation pattern of the dipole antenna can be found in [7,8]. The current distribution and the directivity pattern of a half-wavelength dipole antenna are given in Figure 2-2. In (a), the amplitude of the current distribution is normalized to its maximum, which is at the feed point. The directivity of the radiation pattern in (b) shows a maximum of 2.15 dB. Note that the H-plane pattern is omnidirectional in the x-y plane, and as such it has an infinite number of E planes (all the elevation planes).

FIGURE 2-2 ■
Analytical solution
for a half-wavelength
dipole antenna:
(a) Current
distribution.
(b) E plane directivity
pattern.

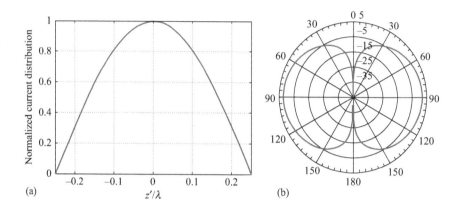

Therefore, when attempting to accurately analyze the radiation pattern, the major concern is correctly approximating the actual current distribution on the dipole. In general, with dipole antennas this requires a full-wave simulation. The next section examines dipole antenna radiation performance using FEKO and compares the current distribution on the wire element with the analytical approximations.

2.3 | DIPOLE ANTENNA FULL-WAVE SIMULATION

2.3.1 Problem Setup

To simulate a finite length dipole antenna in FEKO, we set up the simulation problem in the CADFEKO design environment. First we define the following variables:

- freq = 300e6
- freq_min = 270e6
- freq_max = 330e6
- lambda0 = c0/freq
- dipole_length = 0.5
- dipole_radius = 0.0001

To create these variables, right-click on the variables folder in the main directory and select "Add variable." A new screen will then appear where the selected variable can be defined. A screenshot in which the variable "dipole_length" was defined is given in Figure 2-3.

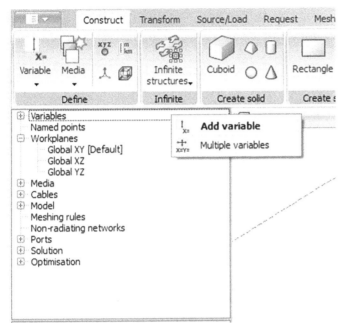

(a)

(b)

FIGURE 2-3 ▪
Defining a variable in
CADFEKO: (a) Add
variable tab.
(b) Defining a
variable.

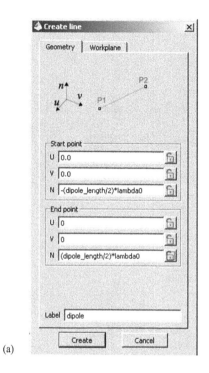

(a)

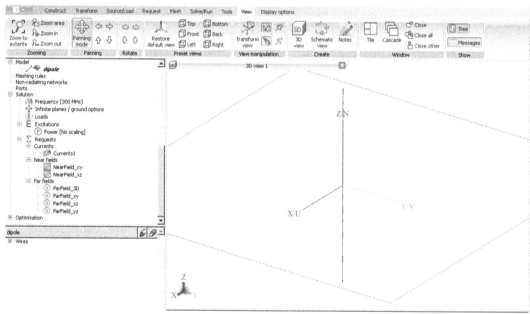

(b)

FIGURE 2-4 ▪ Designing a wire dipole in CADFEKO: (a) Create line menu. (b) Wire line model in FEKO.

Here the dipole antenna is designed for the operating frequency (freq) of 300 MHz. The lower and upper frequencies (freq_min and freq_max) are used for the frequency sweep function. The geometrical parameters of the dipole— the dipole length (dipole_length) and wire radius (dipole_radius)—are both given in terms of the wavelength (lambda0).

Using these variables the design procedure is as follows. In the construct tab, select "line" in the "Create curve" menu. Here we create a line with the start and end coordinates of (0, 0, -dipole_length/2) and (0, 0, dipole_length/2). A snapshot of this is given in Figure 2-4.

Next we define a wire port at the center of the line. In the main menu, right-click on the "Ports" folder and select the "Wire port" option. Then select the wire and place the port at the middle of the wire. A voltage source is then added to the wire port using the default values for the magnitude and phase of the voltage. To define the voltage source, go to the solution directory in the main menu, and under "Excitations" select "Voltage source." Figure 2-5 is a snapshot of the defined voltage source excitation for the dipole.

To simulate the structure we first have to define the solution frequency. Here we set the desired operating frequency (freq) to 300 MHz. For a broadband sweep we can use the lower and upper frequency values. To observe the radiation performance we request the computation of both near and far fields. These results are given in Figure 2-6 for the dipole antenna.

In (a), the total electric field distribution in the x-z plane is shown, where the maximum electric field is observed at the ends of the dipole. The gain pattern of the dipole antenna is shown in (b). It should be noted that in FEKO, reference to gain values in dB and dBi are equivalent. As discussed earlier, the radiation pattern is omnidirectional in the horizontal plane, and all elevation plane patterns (φ-cuts) are identical. These 2D pattern cuts are similar to the analytical pattern shown in Figure 2-2b.

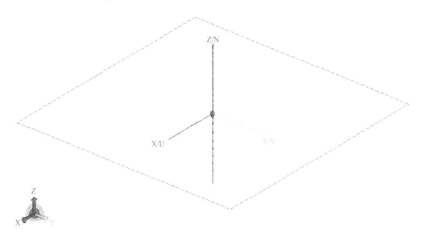

FIGURE 2-5 ■ The constructed dipole antenna geometry in CADFEKO with a voltage source excitation in the middle.

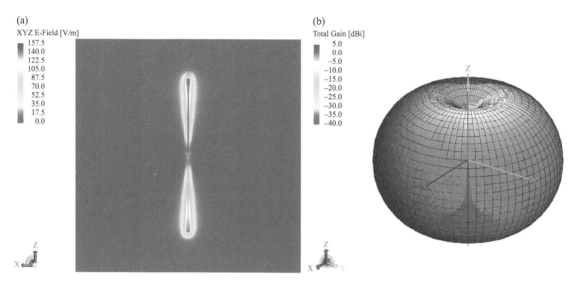

FIGURE 2-6 ■ Radiation performance of a dipole antenna in POSTFEKO: (a) Near electric fields. (b) 3D far electric field pattern.

2.3.2 Parametric Studies

Using the simulation setup given in the previous section, let's study the radiation characteristics of small dipole antennas. We consider two lengths here, $\lambda/20$ and $\lambda/10$, for the dipole. The center design frequency is 300 MHz, and the radius of the wire is 0.0001λ. The amplitude of the current distribution on the dipoles at the center frequency and at the frequency extremes across a 20% band is given in Figure 2-7.

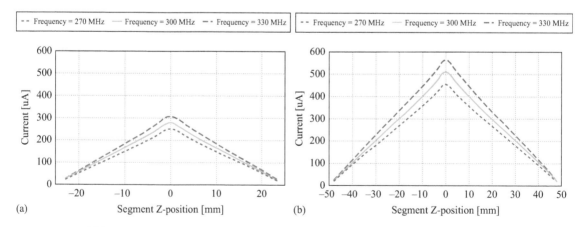

FIGURE 2-7 ■ Current distribution on small dipole antennas: (a) $\lambda/20$. (b) $\lambda/10$.

Indeed, a triangular current distribution can be observed for small wire dipole elements. In addition, for shorter dipoles the magnitude of the current is also smaller and decreases almost monotonously with the dipole length. As the frequency changes, the triangular distribution of the current is still maintained, but the magnitude increases (decreases) as the frequency increases (decreases), since the electrical length increases (decreases).

The corresponding gain patterns of the dipole antennas are given in Figure 2-8, where the maximum gain value is in the order of 1.77 dBi. It should be noted that the maximum scale in both of these figures is 10 dBi. An almost stable radiation pattern is observed for both designs. In general, shorter dipoles can achieve a larger bandwidth, but in practice the main challenge is its feed design.

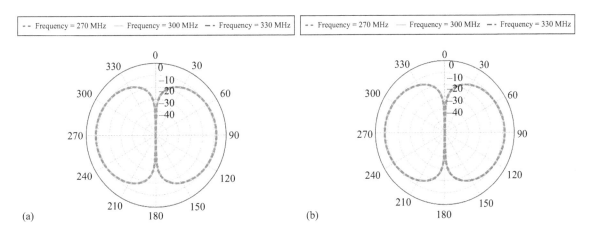

FIGURE 2-8 ■ Gain pattern for small dipole antennas: (a) $\lambda/20$. (b) $\lambda/10$.

Small dipoles generally have a very low gain; therefore, in practice finite length dipoles are usually preferred. Here we study the radiation characteristics of finite length dipoles.

We consider four dipoles with lengths $\lambda/4$, $\lambda/2$, $3\lambda/4$, and λ. Similarly, the center frequency is selected to be 300 MHz, and the radius of the wire is 0.0001λ. The current distribution on the dipoles at the center frequency and at the frequency extremes across a 20% band is given in Figure 2-9.

As the dipole length increases, the triangular current distribution changes into a sinusoidal distribution, where an almost half-cycle is observed for the half-wavelength dipole. To emphasize here the necessity of full-wave simulations for an accurate analysis, the normalized current distribution for half-wavelength and one-wavelength dipoles obtained using the full-wave simulation are compared with the analytical current distribution. These results are given in Figure 2-10.

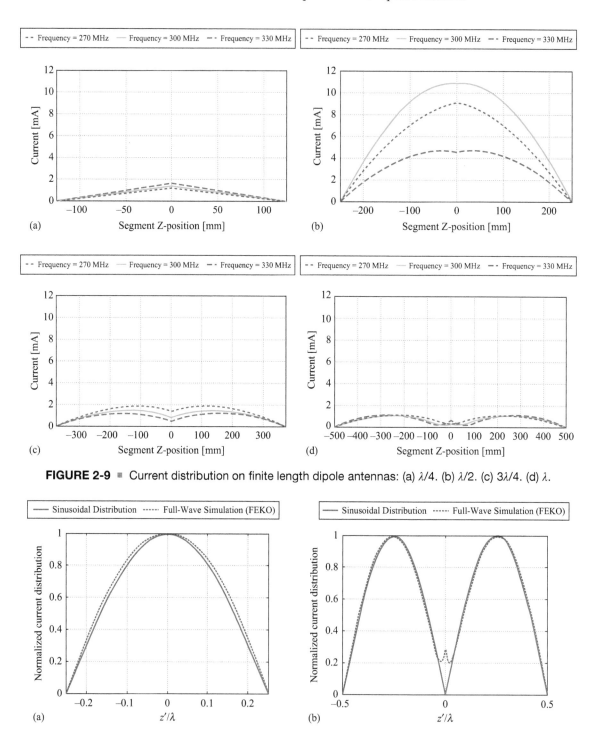

FIGURE 2-9 ■ Current distribution on finite length dipole antennas: (a) $\lambda/4$. (b) $\lambda/2$. (c) $3\lambda/4$. (d) λ.

FIGURE 2-10 ■ Normalized current distribution on a wire dipole antennas: (a) Half-wavelength. (b) One wavelength.

Although the analytical current distribution shows a reasonably close agreement with the full-wave results, some discrepancies are evident. In fact, the agreement between this simple analytical model and full-wave simulation deteriorates as the wire length increases, and typically full-wave simulation is required for an accurate analysis.

For these finite length dipoles, the frequency change results in a stronger variation in the current distribution as the dipole length increases. This corresponds to a greater change in the radiation pattern of larger dipoles as a function of frequency. The corresponding gain pattern of these dipole antennas is given in Figure 2-11. An almost stable radiation pattern is observed for the shorter dipoles, there is a notable difference in the gain pattern when $l = \lambda$.

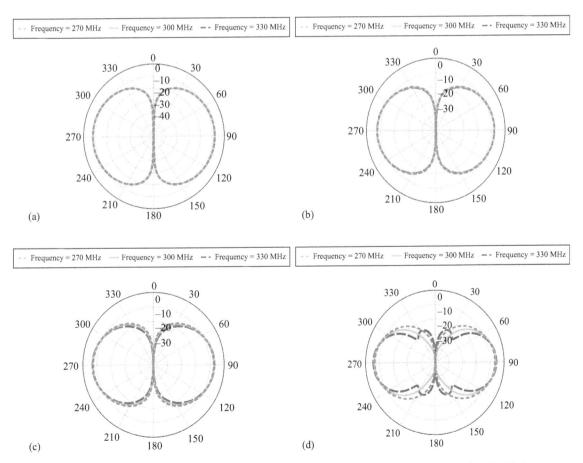

FIGURE 2-11 ▪ Gain pattern for finite length dipole antennas: (a) $\lambda/4$. (b) $\lambda/2$. (c) $3\lambda/4$. (d) λ.

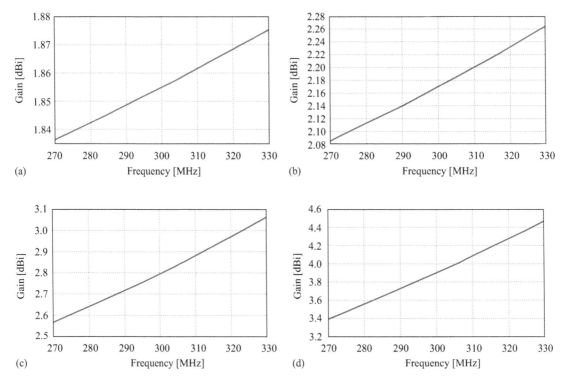

FIGURE 2-12 ■ Gain as a function of frequency for finite length dipole antennas: (a) $\lambda/4$. (b) $\lambda/2$. (c) $3\lambda/4$. (d) λ.

The antenna gain as a function of frequency is given in Figure 2-12. The antenna gain increases with frequency, which is due to the larger electrical size of the dipole. In addition, while all designs show an increase in gain as a function of frequency, this variation in gain increases for larger dipoles. For the case of $l = \lambda$, the change in antenna gain is about 1 dB across this 20% band. It is worth pointing out that in these studies the antenna is assumed to be matched at all frequencies.

For a practical design, where the antenna is fed with a 50 Ω transmission line, matching at the port becomes an important issue. The real and imaginary parts of the impedance at the antenna port are given in Figure 2-13. These results show that the impedance of shorter dipoles (e.g., $\lambda/4$) has a very small real part and a very large imaginary part, while larger dipoles (e.g., $3\lambda/4$) have very large real and imaginary parts. Half-wavelength dipoles, on the other hand, have a practical range of impedance at the port, and as such matching of these dipoles can easily be realized by tuning the dipole length.

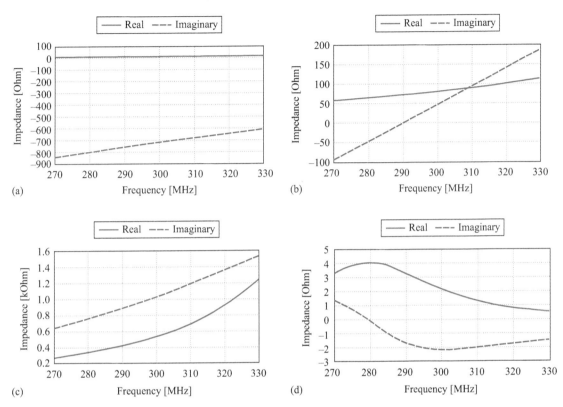

FIGURE 2-13 ■ Input impedance as a function of frequency for finite length dipole antennas: (a) $\lambda/4$. (b) $\lambda/2$. (c) $3\lambda/4$. (d) λ.

2.4 | MONOPOLE ANTENNAS AND FULL-WAVE SIMULATIONS

2.4.1 Infinite Ground Model

A monopole is the top half of a dipole that is placed slightly above a ground plane with the feed point between the ground plane and the lower end of the wire (Figure 2-14).

From antenna theory we know that the radiation pattern in the upper hemisphere of a monopole above a perfect electric ground plane is the same as that of a dipole similarly positioned in free space. The procedure for simulating a monopole antenna in FEKO is almost identical to that for a dipole antenna, as covered in the previous section. Here we will simulate an ultra

FIGURE 2-14 ■
Geometrical
arrangement of a
wire monopole
antenna and the
observation point, P.

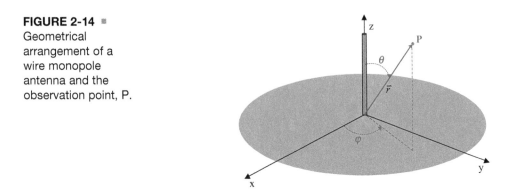

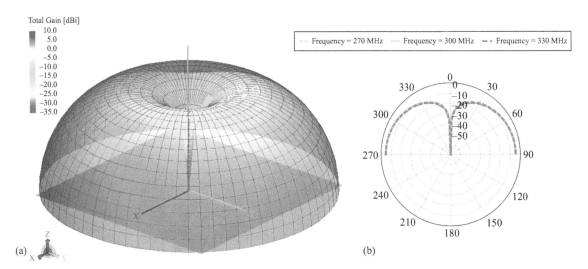

FIGURE 2-15 ■ Radiation pattern of a quarter-wavelength monopole antenna above an infinite ground
plane: (a) 3D pattern at 300 MHz. (b) 2D patterns at three frequencies.

high frequency (UHF) monopole antenna with a length of $\lambda/4$. To simulate
the ground plane for the monopole, in the solution tab, select "infinite
plane/ground" options and "infinite PEC ground" at $z = 0$. When the
infinite ground option is selected, the pattern can be computed only in the top
hemisphere. Thus, for the 3D far-field request, set θ to be from $0°$ to $90°$.
The radiation pattern of this quarter-wavelength monopole antenna is given in
Figure 2-15.

The general pattern shape is quite similar to the half-wavelength dipole
studied earlier. However, the 5.18 dBi maximum gain of the monopole antenna
is 3 dB higher than the 2.17 dBi maximum gain of the dipole antenna.

The current distribution of the monopole antenna is given in Figure 2-16a.
Although the shape of the current distribution closely matches the top half of
the dipole antenna studied earlier, the magnitude is much higher.

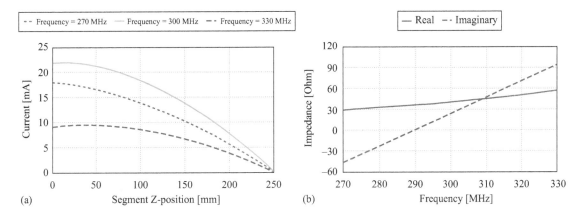

FIGURE 2-16 ■ (a) Current distribution on a monopole antenna. (b) Input impedance as a function of frequency for the monopole antenna.

The input impedance of the antenna as a function of frequency is given in Figure 2-16b. Similarly, whereas the variation of the input impedance of a monopole as a function of frequency is similar to a dipole, the magnitude is much lower. In fact, the input impedance of the monopole is half the input impedance of a dipole antenna.

2.4.2 Finite Ground Model

As discussed earlier, monopole antennas are placed above a ground plane where in theory they are infinite. However, in practical applications, the ground plane usually is of finite size, which alters the radiation characteristics of the monopole. Here we will study the same quarter-wavelength monopole placed above a finite circular ground plane. To model the ground plane in FEKO, we create an ellipse with a radius of 2.5λ. Next we select the face of the ellipse and assign it as a "PEC". To efficiently perform the simulation here with the finite ground plane, we will use the physical optics solution option in FEKO. In the solution tab, select "physical optics (PO) – full ray-tracing." Next select both the monopole and ellipse and unite them. The geometry of this finite monopole antenna, along with the current distribution on the ground plane, is shown in Figure 2-17a. A contour plot showing the current distribution from the top view is given in Figure 2-17b.

As expected, the current is mostly concentrated at the center of the ground plane; thus, the current distribution and input impedance of this finite ground monopole will be similar to the infinite case studied earlier. The current distribution on the monopole and the input impedance as a function of frequency are given in Figure 2-18.

For the current distribution and input impedance, a very close agreement can be observed between the finite (Figure 2-18) and infinite (Figure 2-16)

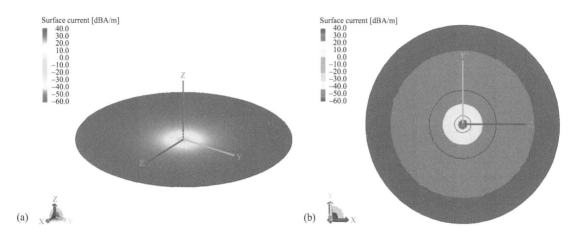

FIGURE 2-17 ▪ The quarter-wavelength monopole antenna above a finite ground plane: (a) 3D model in FEKO. (b) Contour plot of the current distribution on the ground plane.

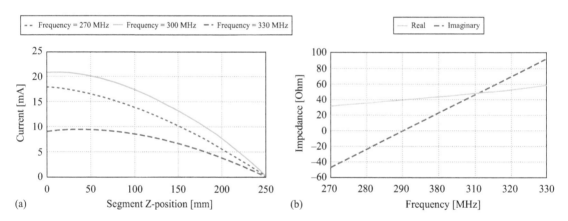

FIGURE 2-18 ▪ (a) Current distribution on a monopole antenna with a 5λ ground plane. (b) Input impedance as a function of frequency for a monopole antenna with a 5λ ground plane.

ground plane cases. But how about the radiation pattern of the monopole? The radiation pattern of this quarter-wavelength monopole antenna is given in Figure 2-19.

It is obvious that while the general pattern shape matches that of the infinite model (Figure 2-15), the radiation pattern is slightly distorted due to the diffractions from the ground plane edge. Nonetheless, the peak gain is almost similar to the infinite case. For monopole antennas with finite ground planes, usually as long as the size of the ground plane is not very small (edge to edge is larger than one wavelength), the main difference between the finite and infinite cases will be a slight distortion in the radiation pattern shape.

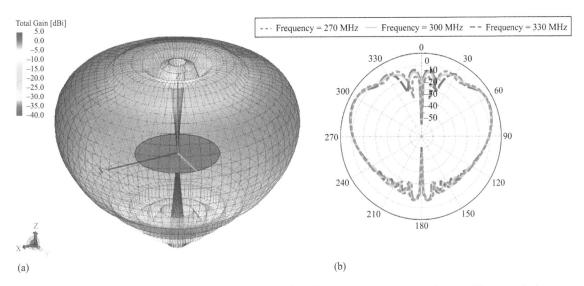

FIGURE 2-19 ■ Radiation pattern of a quarter-wavelength monopole antenna above a 5λ ground plane: (a) 3D pattern at 300 MHz. (b) 2D patterns at three frequencies.

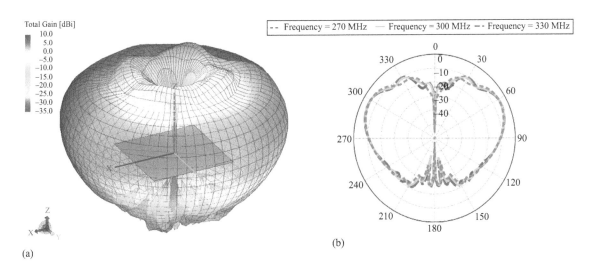

FIGURE 2-20 ■ Radiation pattern of a quarter-wavelength monopole antenna above a 5λ × 5λ square ground plane: (a) 3D pattern at 300 MHz. (b) 2D patterns at three frequencies.

Here let's also study the effect of the ground plane shape on the radiation pattern. The same monopole antenna is now placed above a square ground plane with a dimension of 5λ × 5λ. The radiation patterns for this monopole antenna are given in Figure 2-20.

An almost similar pattern is observed here; however the diffraction from the square shape ground plane results in an asymmetric radiation pattern.

■■■ 2.5 | DIPOLE ANTENNAS AND WIRELESS POWER TRANSFER

As discussed in the previous section, in practice the antenna is fed with a 50 Ω transmission line, so proper matching at the port is of significant importance in the design. Furthermore, accurate tuning of the dipole length is required to ensure that the best matching is obtained at the center design frequency. In this section, we will first design a UHF dipole antenna operating at 300 MHz. In the next stage we will study wireless power transfer using these dipole antennas.

To tune the resonant frequency, the length of the dipole is changed parametrically from 0.48 to 0.49λ. The optimum value is then determined to be 0.4823λ. Here the dipole radius is 0.0001λ. An incident power of 0.1 W is assigned to the 50 Ω port. The reflection coefficient and antenna gain are obtained across the frequency range of 250 to 350 MHz. These results are given in Figure 2-21.

The antenna is well matched at 300 MHz and has an impedance bandwidth ($|S_{11}| < -10$ dB) of about 15 MHz. Note that while the –3 dB gain bandwidth is much larger (about 55 MHz), for a wireless power transfer the major concern is the return loss bandwidth of the transmitter and receiver.

To study wireless power transfer, we consider two dipole elements, placed at a distance (d) from each other. Each dipole antenna has a gain about 2 dB. Both are aligned parallel to the z direction, so theoretically no polarization mismatch will be observed here. The geometry of this simulation in CADFEKO is given in Figure 2-22. The power received by the receiver antenna is given in Figure 2-23 for two cases, $d = \lambda$ and $d = 5\lambda$. For both cases the full-wave simulation results are also compared with the analytical results using the Friis transmission formula [7,8]. When the distance between the two antennas is small, some discrepancy is observed between these results. This is because coupling effects cannot be ignored when the antennas are not in the far-field region from each other, and hence the Friss equation is not accurate in this case. In general, as the separation distance between the antennas increases, a better agreement between analytical and simulation results will be obtained; for the case of $d = 5\lambda$, almost a perfect match is obtained.

■■■ 2.6 | DIPOLE ANTENNA ABOVE A PEC GROUND PLANE

Until now we have considered the radiation characteristics of dipoles radiating into an unbounded medium. In general, the presence of an obstacle, especially when it is near the radiating element, will alter the overall radiation

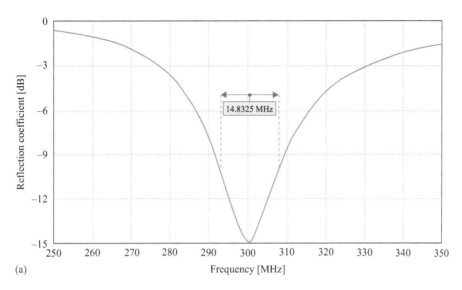

(a)

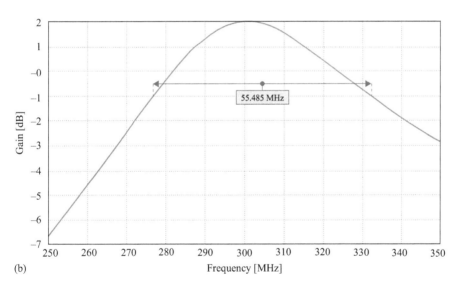

(b)

FIGURE 2-21 ▪
(a) Reflection
coefficient of a
dipole antenna as
a function of
frequency.
(b) Antenna gain as
a function of
frequency.

properties of the antenna system. A very interesting configuration is the dipole antenna when it is placed above an infinite perfect electrical conductor (PEC) ground plane. Theoretical analysis of this configuration is done using image theory [7,8].

Here we will study two cases: the vertical and horizontal half-wavelength dipoles above an infinite PEC ground plane. The geometry of these two configurations is shown in Figure 2-24. The center of the dipole is placed at a distance of one wavelength from the ground plane. Studying the effect of

FIGURE 2-22 ▪
Setup for wireless
power transfer using
two dipole antennas.

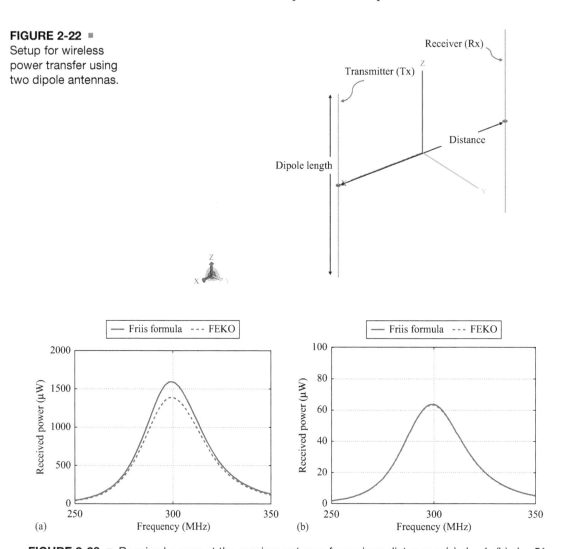

FIGURE 2-23 ▪ Received power at the receiver antenna for various distances: (a) $d = \lambda$. (b) $d = 5\lambda$.

distance to ground plane is left as an exercise for the reader. First let's look at
the current distribution on the wire. For both configurations the current distribution at the center frequency of 300 MHz and two off-center frequencies are
given in Figure 2-25.

Although a small difference is observed in the current distribution of these
two configurations, the general current shape is almost identical. The radiation
pattern, however, is quite different. The radiation pattern in all elevation planes
is identical for the vertical dipole but not for the horizontal dipole. The elevation plane radiation pattern for the vertical dipole is given in Figure 2-26.

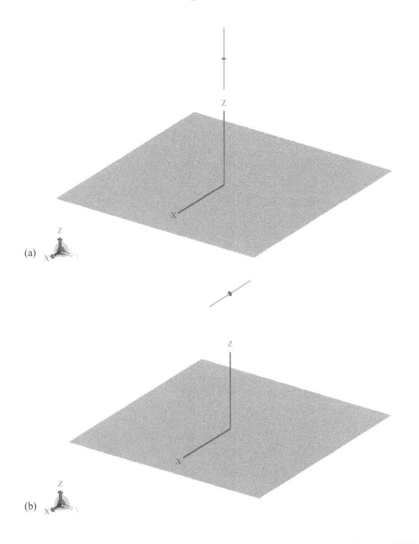

FIGURE 2-24 ▪
Geometry of a dipole
antenna above an
infinite ground plane:
(a) Vertical dipole.
(b) Horizontal dipole.

(a)

(b)

Compared with the dipole in free space (Figure 2-11), a significant difference is now observed in the pattern shape. While a null is still seen in the
broadside direction, several other nulls seem to appear along other elevation
angles. The presence of these nulls is attributed to the existence of the image
source here [7]. Note that if the vertical distance of the dipole to the ground
plane changes, the position of these nulls will change; however, the broadside
null will always be present in this case.

Next we study the horizontal dipole case. The elevation plane radiation
pattern is given in Figure 2-27 for two plane cuts: the x-z and y-z planes. Even
though the dipole is oriented along the x axis, this configuration still has a null
at broadside. At off-center frequencies, however, this null moves away from the
broadside direction. More discussion on this will be given in Chapter 11.

FIGURE 2-25 ▣
Current distribution
on a half-wavelength
dipole antenna
above a PEC ground
plane: (a) Vertical
dipole. (b) Horizontal
dipole.

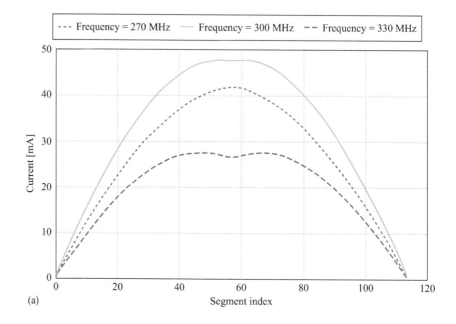

(a)

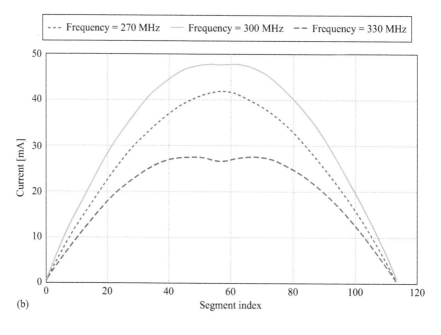

(b)

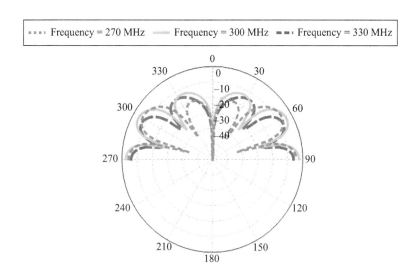

FIGURE 2-26 ▪ Radiation pattern of a vertical half-wavelength dipole antenna above a PEC ground plane.

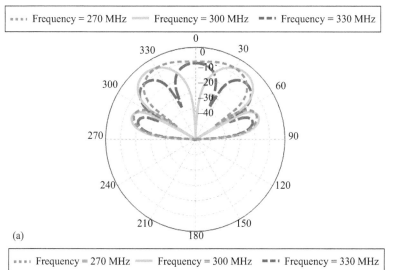

(a)

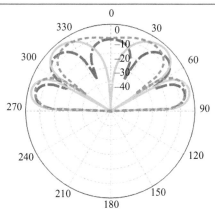

(b)

FIGURE 2-27 ▪ Radiation pattern of a horizontal half-wavelength dipole antenna above a PEC ground plane: (a) $\phi = 0°$. (b) $\phi = 90°$.

▮ 2.7 | DIPOLE ANTENNA NEAR A PEC CYLINDER

In many practical applications, the antenna is placed in a medium where the surrounding objects will distort the radiation pattern of the antenna. Here we will study the radiation performance of the UHF dipole antenna designed in section 2.5 when it is placed near a finite-length PEC cylinder. The geometrical model of the problem is shown in Figure 2-28.

FIGURE 2-28 ▪
A dipole antenna placed near a PEC cylinder.

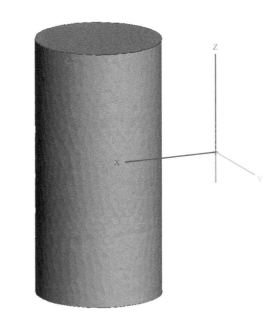

The cylinder is positioned parallel to the dipole antenna and has a radius of 0.5λ and height of 2λ. The center of the cylinder is placed at a distance of λ from the center of the coordinate system; that is, the minimum distance between the cylinder and the dipole is 0.5λ. The current distribution on the cylinder is shown in Figure 2-29. It should be noted that to efficiently perform the simulation, the physical optics solution is used for the cylinder. The presence of this large scattering object implies that the radiation pattern of the dipole antenna will be distorted. The radiation pattern of the antenna is shown in Figure 2-30.

Whereas some notable distortion in the pattern shape of a dipole is observed, the peak gain of the dipole is much larger because the cylinder is working as a reflector, which if designed properly can increase the gain of the antenna. More discussion on reflector antenna designs will be given in the corresponding chapter.

Surface current [mA/m]

36.0
32.0
28.0
24.0
20.0
16.0
12.0
8.0
4.0
0.0

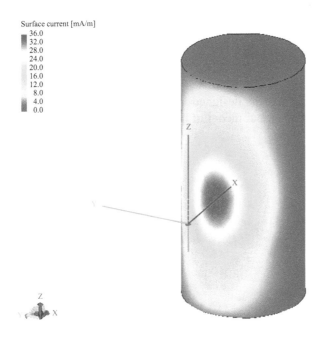

FIGURE 2-29 ▪
Current induced on
the surface of a PEC
cylinder due to
radiation from a
half-wavelength
dipole.

Total Gain

4.5
4.0
3.5
3.0
2.5
2.0
1.5
1.0
0.5
0.0

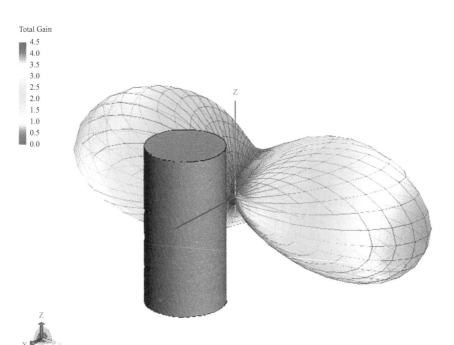

FIGURE 2-30 ▪
Radiation pattern of
a half-wavelength
dipole placed close
to a PEC cylinder.

▮ 2.8 | DIPOLE ANTENNA NEAR A PEC SPHERE

Similar to the previous section, here we will study the radiation performance of the UHF dipole antenna designed in section 2.5 when it is placed near a PEC sphere. The geometrical model of the problem is shown in Figure 2-31.

FIGURE 2-31 ▪
A dipole antenna placed near a PEC sphere.

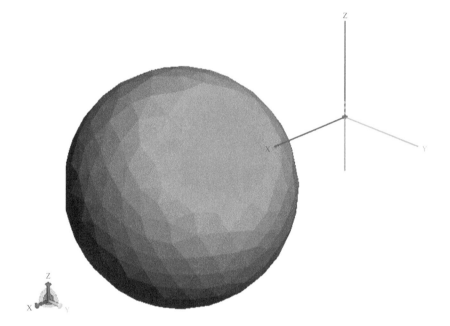

The sphere has a radius of 0.5λ, and the center of the sphere is placed at a distance of λ from the center of the coordinate system; that is, the minimum distance between the sphere and the dipole is 0.5λ. Similarly, we will use the physical optics solver for the PEC sphere. The current distribution on the sphere is shown in Figure 2-32.

The radiation pattern of the dipole antenna will be distorted. The radiation pattern of the antenna is shown in Figure 2-33.

It is also interesting to compare the gain values of the dipole antennas when they are placed close to these obstacles. In the configurations studied in these sections, the maximum dipole gain is obtained with the horizontal dipole when it is placed above an infinite PEC ground. The peak gain in this case is about 4.6 dBi. For the large cylinder case, the peak gain was also about 4.5 dBi, and the sphere configuration had the lowest gain, which was about 3 dBi. Nonetheless, all three configurations achieved a gain higher than the isolated dipole (2.15 dBi) because

Surface current [mA/m]

32.0
28.0
24.0
20.0
16.0
12.0
8.0
4.0
0.0

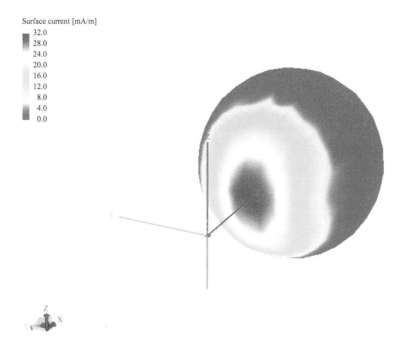

FIGURE 2-32 ▪
Current induced on
the surface of a PEC
sphere due to
radiation from a
half-wavelength
dipole.

Total Gain

3.0
2.7
2.4
2.1
1.8
1.5
1.2
0.9
0.6
0.3
0.0

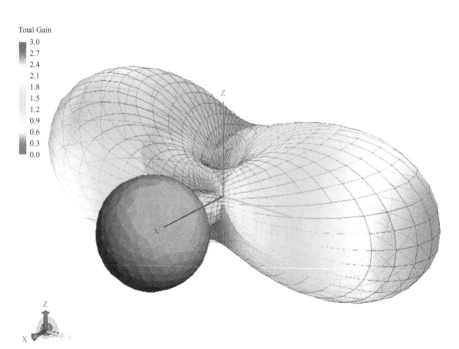

FIGURE 2-33 ▪
Radiation pattern of
a half-wavelength
dipole placed close
to a PEC sphere.

the obstacles act as a reflector, which can increase the gain of the radiating element. More discussion on dipole-fed reflector antennas will be given in Chapter 11.

▮ 2.9 | DIPOLE ANTENNA NEAR A DIELECTRIC SPHERE

In the previous two cases, we studied the radiation characteristics of a dipole antenna when it is placed in the vicinity of metallic objects. In many practical cases, however, the antenna is placed close to dielectric objects. A good example is handheld devices, which are typically held in proximity to the human head when operating.

The human head is usually modeled as a dielectric sphere with layers such as the brain, skull, and skin. For this study, though, we consider only a single-layer model where the material has a dielectric constant of 56.8 (dielectric constant of the human brain) and is placed close to a 900 MHz half-wavelength dipole. The sphere has a radius of 5 cm, and its center is placed at a distance of 7 cm from the half-wavelength dipole; that is, the distance between the outer boundary of the sphere and the dipole is 2 cm. The geometrical model of the problem is shown in Figure 2-34. We use the default solver in FEKO, which employs the method of moments (MoM) and multilevel fast multipole method (MLFMM) surface equivalence principal (SEP), but other solution options are available.

FIGURE 2-34 ▪
A dipole antenna placed near a dielectric sphere.

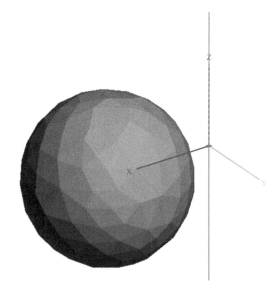

The electric field magnitude inside the sphere is given in Figure 2-35. In this case, a strong field distribution is observed in certain areas inside the sphere. In fact, exposure of brain tissue to electromagnetic fields has raised many concerns about potential adverse health effects [10,11].

The effect of the head on the radiation performance of these devices is also a subject of great interest [10]. The radiation pattern of the dipole antenna is shown in Figure 2-36.

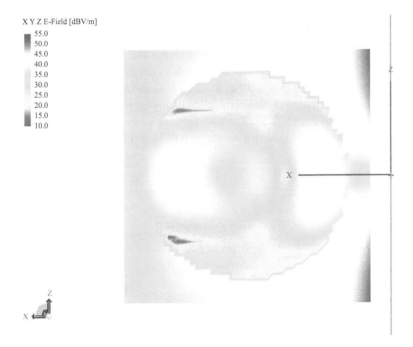

FIGURE 2-35 ▪
Total electric field magnitude inside the dielectric sphere.

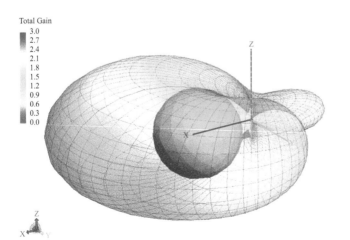

FIGURE 2-36 ▪
Radiation pattern of a half-wavelength dipole placed close to a dielectric sphere.

The radiation pattern of the dipole is highly distorted (compared with the isolated case), and a strong radiated power is going through the sphere.

■ 2.10 | QUASI LOG-PERIODIC DIPOLE ANTENNAS

The dipole antenna has a relatively narrow bandwidth. For example, the UHF dipole antenna studied in the previous section showed a return loss (input impedance) bandwidth less than 15 MHz. To improve the bandwidth of the dipole antenna, one approach is to use lumped elements to improve the matching. In many cases, though, this is not practical so a lumped element can be mimicked using wires [12]. The geometry of the conventional dipole and the quasi log-periodic antenna is shown in Figure 2-37. The two sections of the L shaped wire added to the dipole (Figure 2-37(b)), have a length of $0.05\lambda_0$ and $0.1\lambda_0$, respectively.

FIGURE 2-37 ■
(a) A dipole antenna model in FEKO.
(b) Quasi log-periodic dipole model in FEKO.

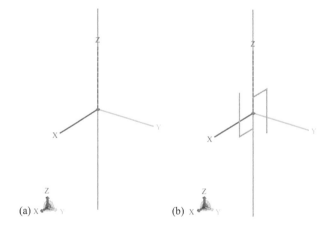

(a)   (b)

In this quasi log-periodic design, two additional mutually coupled stubs are symmetrically connected to the original dipole. They generate more resonance, which tends to cancel the reactive part of the dipole impedance to achieve expanded bandwidth. The impedance of these two dipoles is given in Figure 2-38.

Indeed, the presence of these additional stubs has significantly reduced the imaginary part of the dipole impedance, which would correspond to an increase in the antenna bandwidth. The antenna gain and $|S_{11}|$ as a function of frequency are given in Figure 2-39.

Comparing these results with the conventional dipole reveals that this design can indeed increase the bandwidth of the dipole antenna, where the return loss bandwidth of the quasi log-periodic dipole is raised to almost

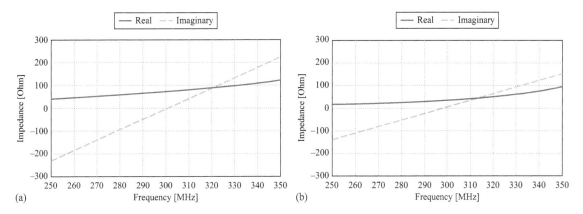

FIGURE 2-38 ▪ Impedance as a function of frequency: (a) Conventional dipole. (b) Quasi log-periodic dipole.

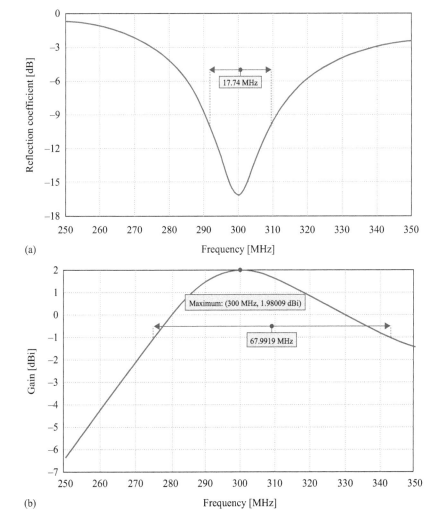

FIGURE 2-39 ▪
(a) Reflection coefficient of a quasi log-periodic dipole antenna as a function of frequency.
(b) Antenna gain as a function of frequency.

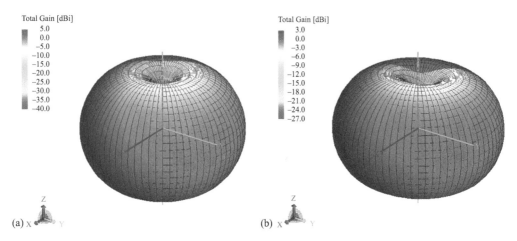

FIGURE 2-40 ▪ Radiation pattern at 300 MHz: (a) Dipole. (b) Quasi log-periodic dipole.

FIGURE 2-41 ▪
Geometrical models
of two four-arm
quasi log-periodic
dipoles.

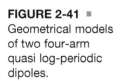

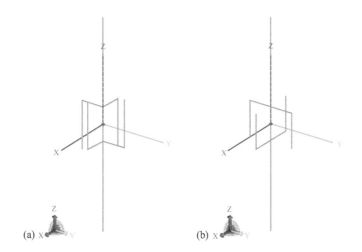

18 MHz. The radiation patterns of the two dipole antennas at 300 MHz are shown in Figure 2-40.

An almost similar omnidirectional radiation pattern is also observed with the quasi log-periodic design. The slight asymmetry observed in the radiation pattern occurs because of the stubs placed along the x axis, which can be corrected by adding additional stubs in the configuration. The geometrical model for two four-arm configurations is shown in Figure 2-41. The analysis of these two configurations is left for readers to experiment with.

EXERCISES

(1) **Effect of Feed Position.** Design a half-wavelength wire dipole antenna with an off-center feed point. Note that in the conventional designs the feed point is at the center of the wire, which results in a symmetric current distribution on the wire. Study the cases where the feed point is located at a distance of $\lambda/8$ and $\lambda/16$ from one end of the wire. Compare the current distribution on the wire and the radiation patterns at the center frequency with the center-fed dipole. Plot the input impedance versus frequency for these two configurations, and discuss your observations.

(2) **Effects of Distance from PEC Ground Plane.** Design a half-wavelength dipole antenna and position it vertically at a distance (d) from the $z = 0$ plane, as shown in Figure P2-1 (a). The center of the dipole should be at $z = d$. Use the infinite ground plane model (section 2.6), and calculate the radiation characteristics of the dipole antenna. Study the effect on radiation performance of the dipole as d changes from $\lambda/2$ to 2λ. What is the difference in the antenna performance compared with the dipole placed in free space? Repeat this exercise when the dipole is placed parallel to the ground plane as shown in Figure P2-1 (b).

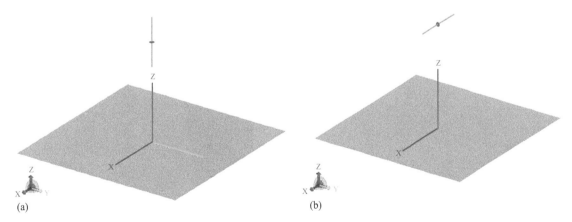

(a)   (b)

FIGURE P2-1 ▪ A dipole antenna placed above an infinite ground plane: (a) vertical dipole, (b) horizontal dipole.

(3) **Dipoles and Monopoles.** Using the infinite ground plane model, calculate the input impedance and gain of monopole antennas with lengths of $\lambda/8$, $\lambda/4$, $\lambda/2$, and λ as a function of frequency with center frequency being set to 300 MHz. Compare your results with Figure 2-10 and Figure 2-11 for dipole antennas with twice the size. What is the relationship between the input impedance and gain of a monopole antenna with a dipole antenna twice its size?

(4) ***Wireless Power Transfer.*** Repeat the study done is section 2.5, with two monopole antennas placed above an infinite ground plane. Compared with the dipole case, does the received power at the port increase? Explain your answer.

(5) ***Dipole Placed Near a PEC Disc.*** Create a circular PEC disc with a diameter of 5λ. Use the PO solution for the disc. Next, place a half-wavelength dipole antenna parallel to the plane of the disc at distance (d). The geometry of this problem in CADFEKO is given in Figure P2-2. Change d from λ to 5λ and observe the gain patterns for this antenna. Does this antenna have a higher gain than a dipole radiating in free space? At what distance does this antenna achieve its maximum gain?

FIGURE P2-2 ■
A dipole antenna placed parallel to a circular PEC disc.

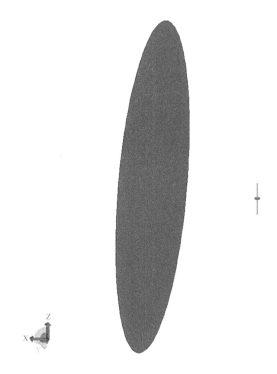

Wire Loop Antennas

Chapter Outline

3.1 | INTRODUCTION

This chapter introduces another fundamental antenna configuration: the loop antenna. This simple, inexpensive, and very versatile antenna takes many different forms, such as a rectangle, square, triangle, ellipse, and circle. Because of the simplicity in analysis and construction, the circular loop is the most popular and has received the widest attention. In this chapter, the effect of loop circumference on the current distribution and radiation performance of the antenna is illustrated using several examples.

3.2 | SMALL AND LARGE LOOP ANTENNAS

Loop antennas are usually classified into two categories: electrically small and electrically large. Electrically small antennas are those whose overall length (circumference) is usually less than 1/10 of a wavelength ($C < \lambda/10$). A small loop is equivalent to an infinitesimal magnetic dipole whose axis is perpendicular to the plane of the loop. That is, the fields radiated by an electrically small circular or square loop are of the same mathematical form as those radiated by an infinitesimal magnetic dipole. Loop antennas with electrically small circumferences or perimeters have small radiation resistances that are usually smaller than their loss resistances. Thus, they are very poor radiators, and they are seldom employed for transmission in radio communication. When they are used in any application like this, it is usually in the receiving mode, such as in portable radios and pagers, where antenna efficiency is not as important as the signal-to-noise ratio. They are also used as probes for field measurements and as directional antennas for radio wave navigation. The field pattern of electrically small loop antennas of any shape (e.g., circular, elliptical, rectangular, square) is similar to that of an infinitesimal electric dipole with a null perpendicular to the plane of the loop and with its maximum along the plane of the loop.

On the other hand, electrically large loops are those whose circumference is equal or larger than the free-space wavelength ($C \geq \lambda$). As the overall length of the loop increases and its circumference approach one free-space wavelength, the maximum of the pattern shifts from the plane of the loop to the axis of the loop, which is perpendicular to its plane. The radiation resistance of the loop can be increased and made comparable to the characteristic impedance of practical transmission lines by increasing (electrically) its perimeter or the number of turns. Electrically large loops are used primarily in directional arrays, such as in helical antennas, Yagi-Uda arrays, and quad arrays (see Chapter 7). For these and other similar applications, the maximum radiation is directed toward the axis of the loop, forming an end-fire antenna. To achieve such directional pattern characteristics, the circumference (perimeter) of the loop should be about one free-space wavelength. The proper phasing between turns enhances the overall directional properties.

A loop antenna can be used as a single element or in array configurations. The mounting orientation of the loop element and the array configuration will determine its overall pattern and radiation characteristics. Most of the applications of loop antennas are in the high frequency (HF; 3–30 MHz), very high frequency (VHF; 30–300 MHz), and ultra high frequency (UHF; 300–3,000 MHz) bands. However, when used as field probes, they find applications even in the microwave frequency range. Figure 3-1 shows a circular loop positioned symmetrically around the origin of the coordinate system with its normal oriented along the z axis.

To calculate the fields radiated by the loop antenna, we use the same procedure as that for the linear dipole in Chapter 2. That is, we start with equation (2-1) and determine the vector potential function (**A**) to compute the far-field radiation pattern. However, like before the challenge is to define a practical mathematical model for the current distribution on the wire. In the simplified analytical models, the current on the loop is assumed to be constant (I_0). This simple approximation is quite valid for small loop antennas but not for large loops, as shown later on in this chapter. Here we will briefly outline

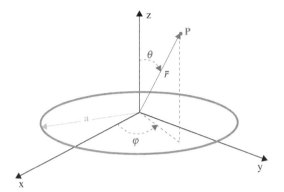

FIGURE 3-1 ▪
Geometrical
arrangement of a
wire loop antenna
and the observation
point, P.

the procedure for calculating the pattern of a small loop antenna with constant current, and readers are referred to [7] for analytical methods for large loops.

For a small loop antenna oriented as shown in Figure 3-1, the ϕ-component of the vector potential is given by [7]

$$A_\phi = \frac{a\mu I_0}{4\pi} \int\limits_0^{2\pi} \cos \phi' \frac{e^{-jk \sqrt{r^2+a^2-2ar \sin \theta \cos \phi'}}}{\sqrt{r^2 + a^2 - 2ar \sin \theta \cos \phi'}} d\phi' \qquad (3\text{-}1)$$

where a is the radius of the loop antenna, and μ is the permeability of the surrounding medium. This integration is too complex to be solved analytically, but it can be simplified by considering only the first two terms of the Maclaurin series for the integrand. Once the vector potential is derived, the radiation pattern is computed as described in [7,8].

3.3 | CIRCULAR LOOP ANTENNAS

3.3.1 Problem Setup

To simulate the loop antenna in FEKO, we set up the simulation problem in the CADFEKO design environment. As an example, first we define the following variables:

- freq = 300e6
- freq_min = 270e6
- freq_max = 330e6
- lambda0 = c0/freq
- loop_C = 0.1
- wire_radius = 0.0001

A screenshot of the variables is shown in Figure 3-2. Here the loop antenna is designed for the operating frequency (freq) of 300 MHz. The lower (freq_min) and upper (freq_max) frequencies are used for the frequency sweep analysis. The loop circumference (loop_C) and wire radius (wire_radius), that is, the geometrical parameters of the loop, are both given in terms of the wavelength (lambda).

FIGURE 3-2 ■ The variables defined in the simulation.

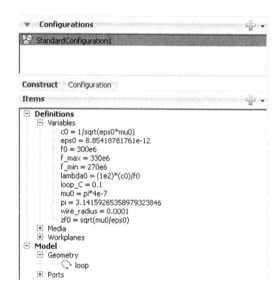

Using these variables, the design procedure is as follows. In the construct tab, select "Elliptic arc" in the "Create curve" menu. Here we create an arc with the center located at (0, 0, 0), and a radius of "loop_C/(2*pi)." A snapshot of this is given in Figure 3-3.

Similar to the dipole case, we next need to define a wire port, so we need to select the segment position on the wire. Here we choose the starting point, which is located at $x = a, y = 0, z = 0$. A screenshot of this is given in Figure 3-4.

A voltage source is then added to the wire port using the default values for the magnitude and phase of the voltage. The geometry of a loop antenna with a loop circumference of one wavelength, i.e. loop_C = 1, along with the feed port is given in Figure 3-5.

To simulate the structure we first define the solution frequency. Here we set "freq" to 300 MHz. For a broadband sweep we can use the lower and upper frequency values as defined previously. To observe the radiation performance we request the computation of both near and far fields. These results are given in Figure 3-6 for the circular loop antenna at 300 MHz.

For this large loop (one wavelength circumference), the pattern is not perfectly symmetric because of the asymmetric current distribution. More discussion on the currents on the loop antenna is given in the next section.

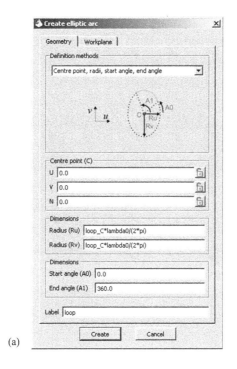

(a)

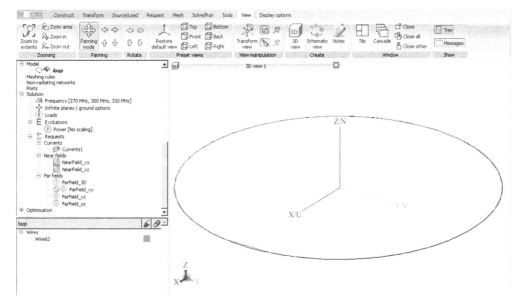

(b)

FIGURE 3-3 ▪ Designing a wire loop in CADFEKO: (a) Create ellipse tab. (b) Wire loop model in FEKO.

FIGURE 3-4 ▪
Creating a wire port
for a loop antenna in
CADFEKO.

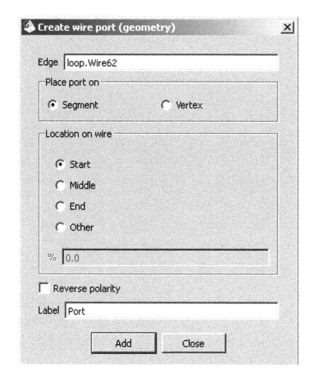

FIGURE 3-5 ▪
Geometry of a
loop antenna in
CADFEKO.

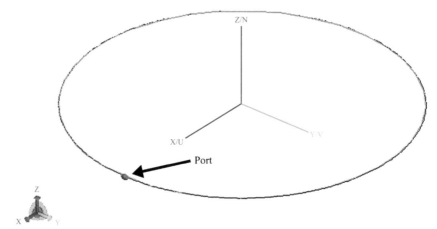

3.3.2 Parametric Studies

Using the simulation setup given in the previous section, now let's examine the
radiation characteristics of electrically small loop antennas. We consider two
cases here for the circumference of the loop: $\lambda/20$ and $\lambda/10$. The design frequency is 300 MHz, and the radius of the perfect electric conductor (PEC) wire

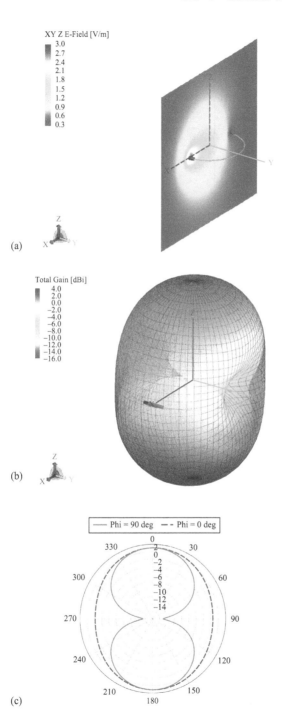

FIGURE 3-6 Radiation performance of a loop antenna with a circumference of one wavelength in POSTFEKO: (a) Near-field distribution in the y-z plane. (b) 3D gain pattern. (c) Gain radiation patterns in y-z and x-z planes.

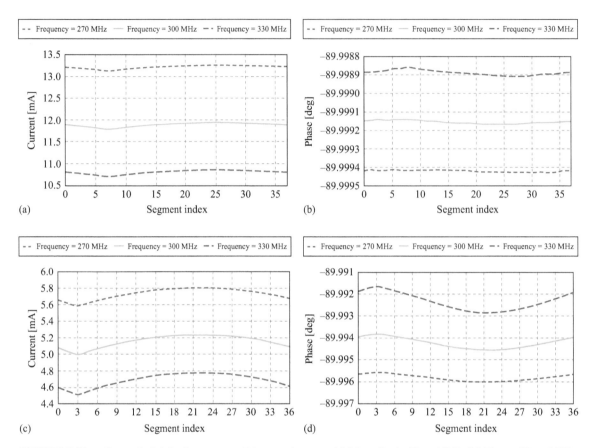

FIGURE 3-7 ■ Current distribution on small loop antennas: (a) Magnitude ($C = \lambda/20$). (b) Phase ($C = \lambda/20$). (c) Magnitude ($C = \lambda/10$). (d) Phase ($C = \lambda/10$).

is 0.0001λ. The current distribution on the loops versus the segment index at the center frequency and at the frequency extremes across a 20% band is shown in Figure 3-7.

For small loops, an almost constant amplitude of the current is observed on the wire. However, the smaller loop shows much less current variations compared with the larger design. Indeed, as the size of the loop increases, a null seems to be forming on the opposite side of the feed point. As the frequency changes within the 20% band, the shape of the current distribution is still maintained, but the magnitude decreases (increases) as the frequency increases (decreases).

The gain pattern of the loop antennas is given in Figure 3-8, in which an almost stable radiation pattern is observed for both designs. Typically, small loops can achieve a larger bandwidth, but in practice the main challenge is the

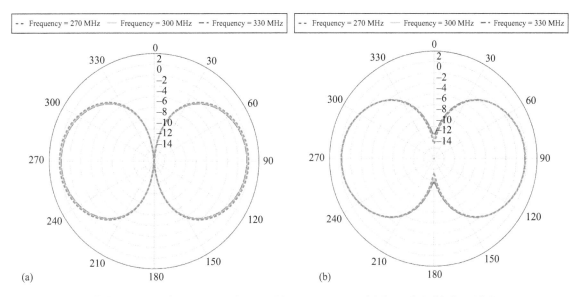

FIGURE 3-8 ▪ Gain pattern for small loop antennas: (a) $C = \lambda/20$. (b) $C = \lambda/10$.

feed design. Furthermore, a very similar radiation pattern as the dipole antenna is observed with the loop antenna. It is worthwhile to point out here that the radiation pattern of small loops is azimuthally symmetric.

As discussed earlier, small loops generally have a very low gain, so in practice large loops are usually preferred. Here we study the radiation characteristics of electrically large loop antennas using four loop circumferences: $\lambda/2$, λ, $3\lambda/2$, and 2λ. The design frequency remains 300 MHz and the radius of the wire 0.0001λ. The current distribution on the loops at the center frequency and at the frequency extremes across a 20% band is given in Figure 3-9.

As the circumference of the loop increases, the almost constant current distribution of the small loop changes into a somewhat asymmetric sinusoidal distribution. Moreover, a significant variation of phase is also observed on the wires. From an analysis viewpoint, simple classical approaches [7,8] that assume a constant current on the loop will show a very high error for large loops. Higher order approximations for the current distribution can improve the accuracy of the results. However, in general the analytical formulations for loop antennas are much more complicated than the wire dipoles, and in most cases for an accurate analysis full-wave simulations are required.

Comparing the current for these four loop sizes shows that in all cases some nulls will be generated on the wire but that the position of the nulls depends on the size of the loop. For the loop antennas with 0.5λ and 1.5λ circumference, the current distribution is asymmetric. As such the beam direction is expected to be tilted. On the other hand, for the loop antennas with λ and 2λ circumference,

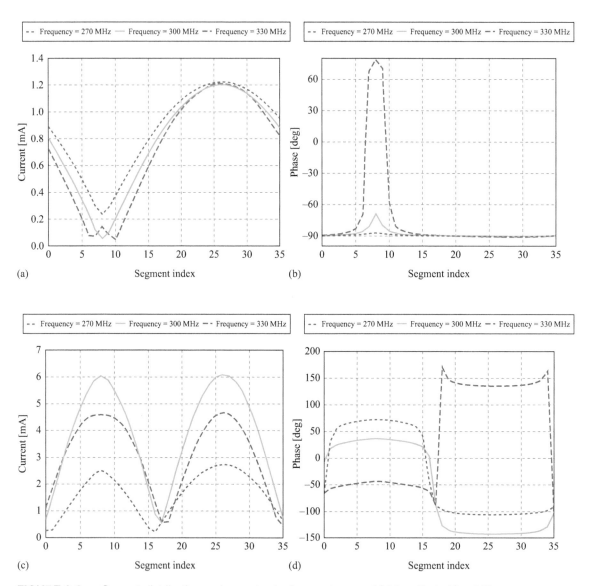

FIGURE 3-9 ■ Current distribution on large circular loop antennas: (a) Magnitude ($C = \lambda/2$).
(b) Phase ($C = \lambda/2$). (c) Magnitude ($C = \lambda$). (d) Phase ($C = \lambda$). (e) Magnitude ($C = 3\lambda/2$). (f) Phase ($C = 3\lambda/2$).
(g) Magnitude ($C = 2\lambda$). (h) Phase ($C = 2\lambda$).

the current distribution is almost symmetric, thus resulting in a more applicable
radiation performance. For these large loops, the frequency change results in
significant variations in the current distribution. This corresponds to a greater
change in the radiation pattern of large loops as a function of frequency, hence
limiting their bandwidth. The gain pattern of these large loop antennas is given in

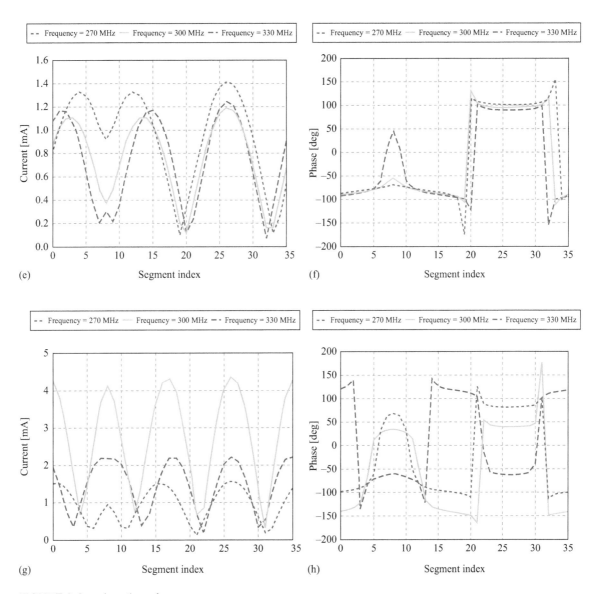

FIGURE 3-9 ▪ *(continued)*

Figure 3-10, where a notable difference is observed in the gain patterns at different frequencies.

Unlike dipole antennas, the mainbeam direction of loop antennas is a function of the loop size. Therefore, it is meaningless to compare the gain of these loop antennas as a function of frequency. However, for a practical design, where the antenna is fed with a 50 Ω transmission line, matching the antenna at

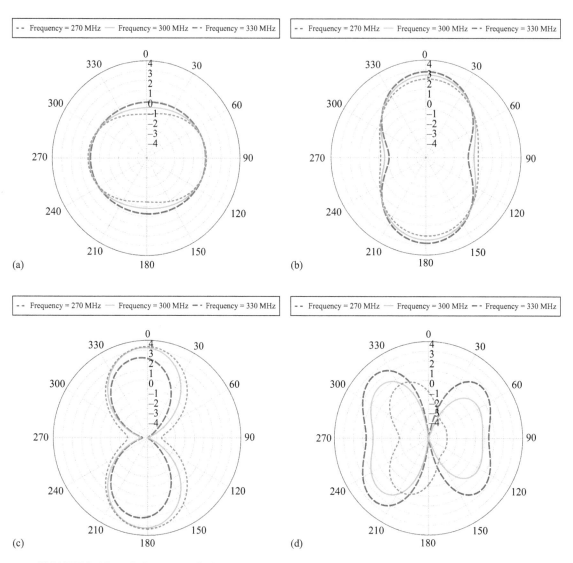

FIGURE 3-10 ▪ Gain pattern for loop antennas: (a) $C = \lambda/2$. (b) $C = \lambda$. (c) $C = 3\lambda/2$. (d) $C = 2\lambda$.

the port is a major issue. The real and imaginary parts of the impedance at the antenna port are given in Figure 3-11. These results show that for loop antennas with a circumference of 0.5λ or 1.5λ, both real and imaginary parts of the impedance are very large, making them very impractical. On the other hand, loop antennas with a circumference of λ or 2λ show a rather acceptable performance. One major disadvantage, though, is the strong increase of the imaginary part of impedance as a function of frequency, which essentially means that the loop

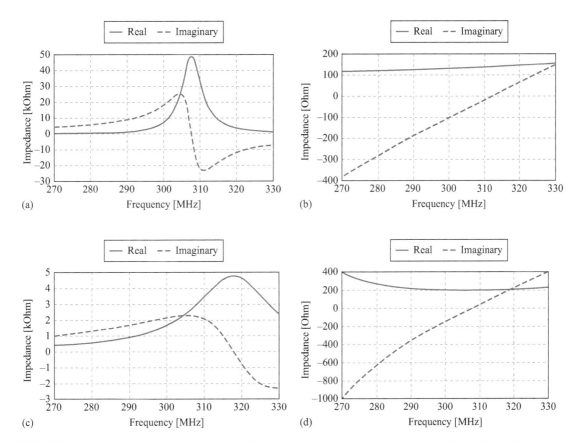

FIGURE 3-11 ▪ Impedance as a function of frequency for large loop antennas. (a) $C = \lambda/2$. (b) $C = \lambda$. (c) $C = 3\lambda/2$. (d) $C = 2\lambda$.

will be very narrow band. Note that the real part is almost constant as a function of frequency.

3.4 | SQUARE LOOP ANTENNAS

3.4.1 Problem Setup

Similar to the circular loop, we first set up the simulation problem in the CADFEKO design environment by defining the following variables:

- freq = 300e6
- freq_min = 270e6
- freq_max = 330e6
- lambda = c0/freq

- loop_circumference $= 0.5$
- wire_radius $= 0.0001$

Using these variables, the design procedure is as follows. First we create a poly line by defining five corner points for the square. The fifth point is basically the initial (or first) point, but it is necessary to be redefined to close the loop. Here we assume that the wire is a PEC. Next we define a wire port at the center of the wire. A voltage source is then added to the wire port using the default values for the magnitude and phase of the voltage. The geometry of the structure is given in Figure 3-12 with the port placed at the intersection of the positive x axis with one of the loop arms.

FIGURE 3-12 ■
Geometry of a square loop antenna in CADFEKO.

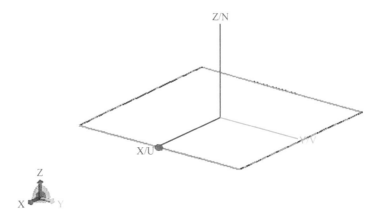

To observe the radiation performance here, we request the computation of both near and far fields. The results are given in Figure 3-13 for this square loop antenna. Similarly, for this large loop (one wavelength circumference), the pattern is not perfectly symmetric because of asymmetric current distribution. More discussion on the currents on the loop antenna is given in the next section.

3.4.2 Parametric Studies

As discussed earlier, the radiation performance of a small loop is independent of its shape; however, for a large loop the current distribution is quite different depending on the loop geometry. Here we study the radiation characteristics of four electrically large square loop antennas with loop circumference $\lambda/2$, λ, $3\lambda/2$, and 2λ. The design frequency remains 300 MHz and the wire radius 0.0001λ. The current distribution on the loops at the center frequency and at the frequency extremes across a 20% band is given in Figure 3-14.

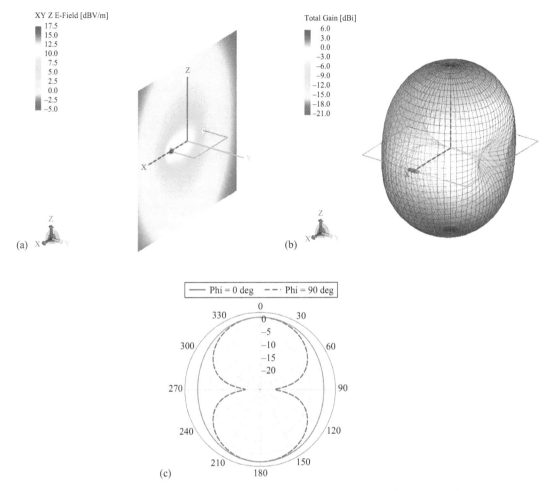

FIGURE 3-13 ▪ Radiation performance of a loop antenna in POSTFEKO: (a) Near-field distribution in the y-z plane. (b) 3D gain pattern. (c) Gain radiation patterns in the x-z and y-z planes.

Like with the large circular loops, in this case some nulls in the current distribution will be generated on the wire. However, the position of the nulls depends on the size of the loop. For the loop antennas with 0.5 and 1.5λ circumference, the current distribution is asymmetric, so the beam direction is expected to be tilted. On the other hand, for the loop antennas with λ and 2λ circumference, the current distribution is almost symmetric, thus resulting in a more applicable radiation performance. The gain pattern of these large loop antennas is given in Figure 3-15, where a notable difference is observed in the gain patterns at different frequencies.

The real and imaginary part of the impedance at the antenna port is given in Figure 3-16. These results show that the performance of the square loop antennas

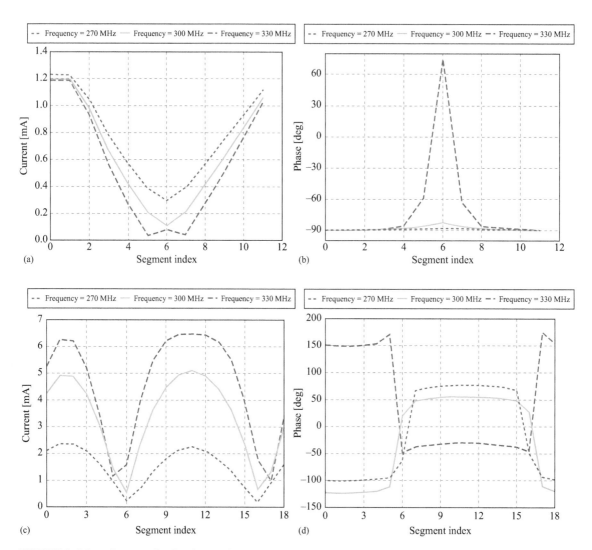

FIGURE 3-14 ■ Current distribution on large square loop antennas: (a) Magnitude ($C = \lambda/2$).
(b) Phase ($C = \lambda/2$). (c) Magnitude ($C = \lambda$). (d) Phase ($C = \lambda$). (e) Magnitude ($C = 3\lambda/2$). (f) Phase ($C = 3\lambda/2$).
(g) Magnitude ($C = 2\lambda$). (h) Phase ($C = 2\lambda$).

in terms of input impedance is similar to circular loops; that is, with a circumference of 0.5λ or 1.5λ, both real and imaginary parts of the impedance are
very large, making them very impractical. On the other hand, loop antennas with a
circumference of λ or 2λ show a rather acceptable performance. Again, the major
disadvantage is the strong increase in the imaginary part of impedance as a function of frequency, which essentially means that the loop will be very narrowband.

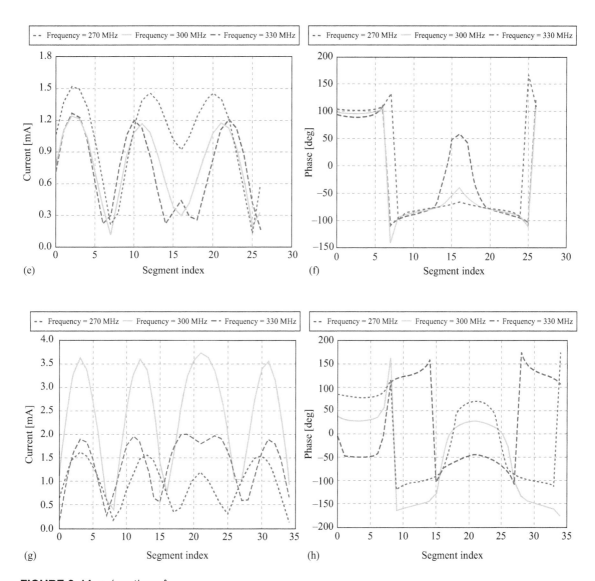

FIGURE 3-14 ▪ (continued)

It is interesting to point out that the radiation performance of the square loop antennas could be different depending on the feed position because unlike the circular loop the square configuration is not azimuthally symmetric. Here we will study the same loop configuration when it is excited at the junction between two sides of the square loop. The geometrical model of this configuration is given in Figure 3-17.

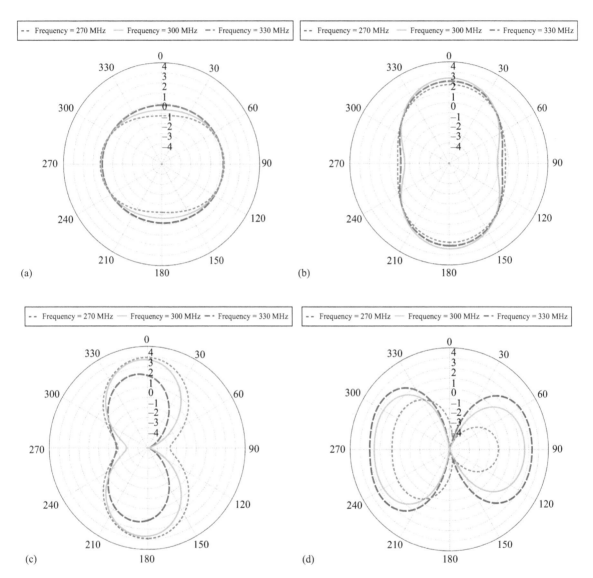

FIGURE 3-15 ■ Gain patterns in the xz-plane for square loop antennas: (a) $C = \lambda/2$. (b) $C = \lambda$. (c) $C = 3\lambda/2$. (d) $C = 2\lambda$.

While the feed is placed at the junction of two wires, the direction of the input current can be specified for only one of the wires. Here we have selected the wire along the y direction. A zoomed view of this corner-fed loop port is shown in Figure 3-18.

Similarly we consider four cases for the loop circumference: $\lambda/2$, λ, $3\lambda/2$, and 2λ. The design frequency is 300 MHz and the wire radius 0.0001λ.

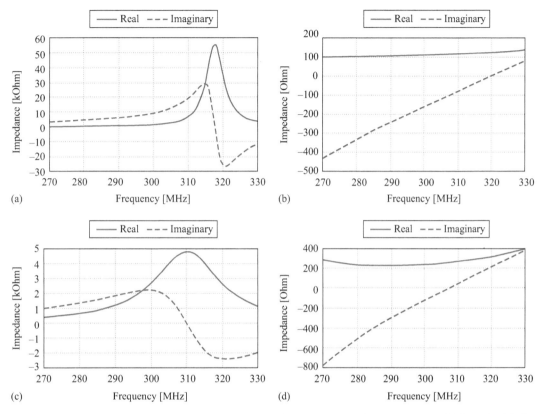

FIGURE 3-16 ■ Impedance as a function of frequency for large square loop antennas: (a) $C = \lambda/2$. (b) $C = \lambda$. (c) $C = 3\lambda/2$. (d) $C = 2\lambda$.

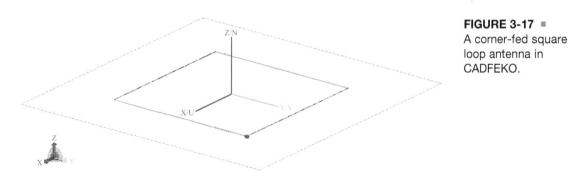

FIGURE 3-17 ■ A corner-fed square loop antenna in CADFEKO.

The current distribution on the loops at the center frequency and at the frequency extremes across a 20% band is given in Figure 3-19.

Comparing this figure with Figure 3-13 reveals that the magnitudes of the current distribution on the wires are similar. However, because of the difference

FIGURE 3-18 ⬚
Zoomed view of the
corner-fed loop port
in POSTFEKO.

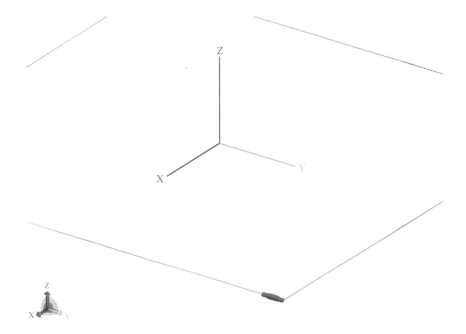

FIGURE 3-18 ⬚
Zoomed view of the corner-fed loop port in POSTFEKO.

in current phase, the radiation patterns will be quite different. The radiation patterns for these four cases are illustrated in Figure 3-20. For larger loops in particular, the radiation patterns are significantly different from the center-fed cases in Figure 3-15. The real and imaginary parts of the impedance at the antenna port is given in Figure 3-21, where a similar observation on the bandwidth performance of these designs can be made.

3.5 | TRIANGULAR LOOP ANTENNAS

3.5.1 Problem Setup

Another interesting configuration is the triangular loop antenna. Similar to the square configuration, this geometry also does not have azimuthal symmetry. We first set up the simulation problem in the CADFEKO design environment by defining the following variables:

- freq = 300e6
- freq_min = 270e6
- freq_max = 330e6
- lambda = c0/freq
- loop_circumference = 0.5
- wire_radius = 0.0001

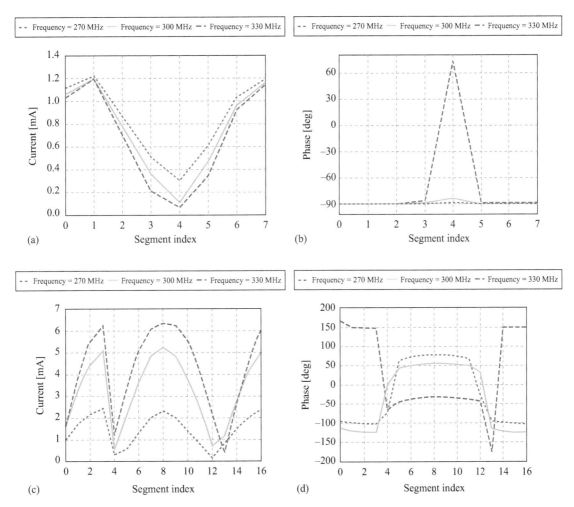

FIGURE 3-19 ▪ Current distribution on corner-fed large square loop antennas: (a) Magnitude ($C = \lambda/2$). (b) Phase ($C = \lambda/2$). (c) Magnitude ($C = \lambda$). (d) Phase ($C = \lambda$). (e) Magnitude ($C = 3\lambda/2$). (f) Phase ($C = 3\lambda/2$). (g) Magnitude ($C = 2\lambda$). (h) Phase ($C = 2\lambda$).

Using these variables, the design procedure is as follows. First we create a poly line by defining four corner points for the triangle. Next we delineate a wire port at the center of one of the wires. Here we assume that the wire is a PEC. A voltage source is then added to the wire port using the default values for the magnitude and phase of the voltage. The geometry of the equilateral triangular loop antenna in CADFEKO is given in Figure 3-22.

To observe the radiation performance, we request the computation of both near and far fields. These results are given in Figure 3-23 for a center-fed

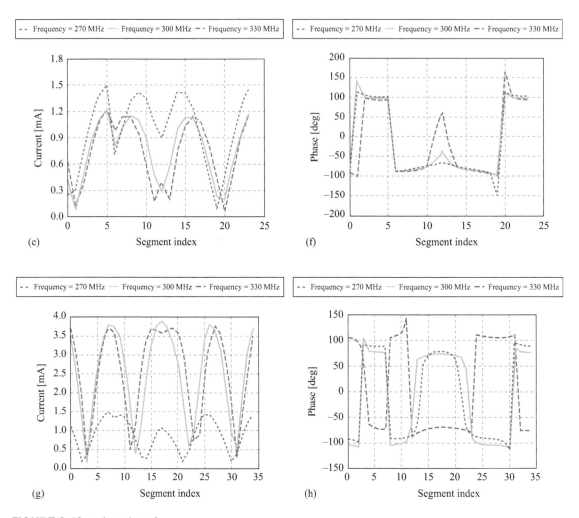

FIGURE 3-19 ▪ (*continued*)

triangular loop with a circumference of one wavelength. For this configuration, the pattern is not azimuthally symmetric; therefore, it would be advantageous to observe the pattern in both the x-z and y-z planes.

3.5.2 Parametric Studies

Here we study the radiation characteristics of electrically large triangular loop antennas using four loop circumferences: $\lambda/2$, λ, $3\lambda/2$, and 2λ. The design frequency is 300 MHz and the wire radius 0.0001λ. First we will study the

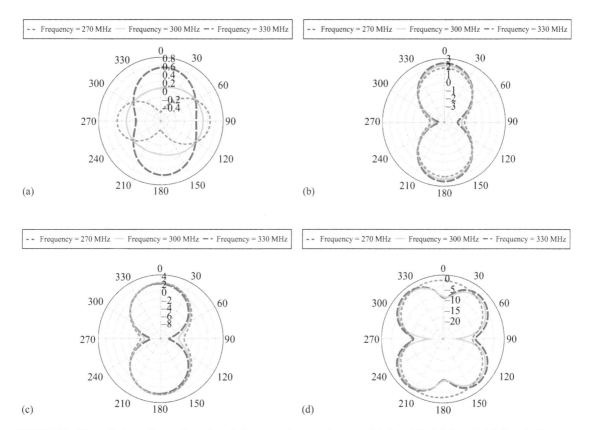

FIGURE 3-20 ▪ Gain patterns for edge-fed square loop antennas. (a) $C = \lambda/2$. (b) $C = \lambda$. (c) $C = 3\lambda/2$. (d) $C = 2\lambda$.

center-fed cases. The current distribution on the loops and the 2D radiation patterns at the center frequency and at the frequency extremes across a 20% band are given in Figure 3-24, Figure 3-25, and Figure 3-26.

Like the other loop configurations we studied earlier, some nulls in the current distribution will be generated on the wire. However, the position of the nulls depends on the size of the loop. For the radiation patterns, for this orientation a symmetric radiation pattern will be observed only in the y-z plane. The real and imaginary parts of the impedance at the antenna port is illustrated in Figure 3-27. A similar observation, as in the previous cases, can be made here where loop antennas with circumference of 0.5λ or 1.5λ have a very large impedance (both real and imaginary parts), which makes them very impractical. On the other hand, loop antennas with a circumference of λ or 2λ show a rather acceptable performance.

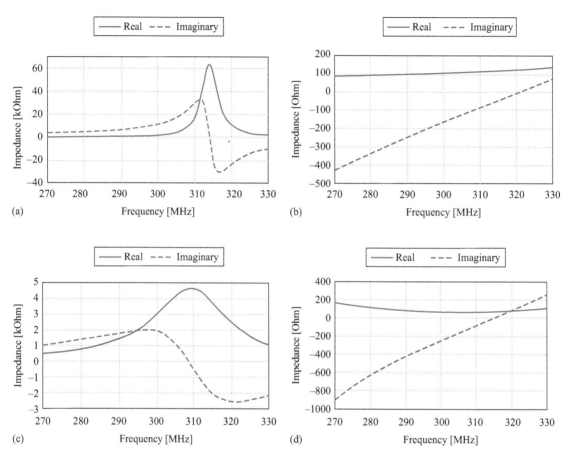

FIGURE 3-21 ▪ Impedance as a function of frequency for edge-fed square loop antennas: (a) $C = \lambda/2$. (b) $C = \lambda$. (c) $C = 3\lambda/2$. (d) $C = 2\lambda$.

FIGURE 3-22 ▪
Geometry of an
equilateral triangular
loop antenna.

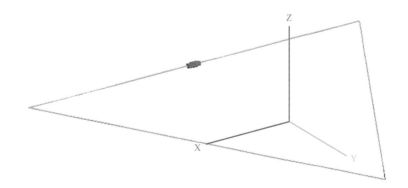

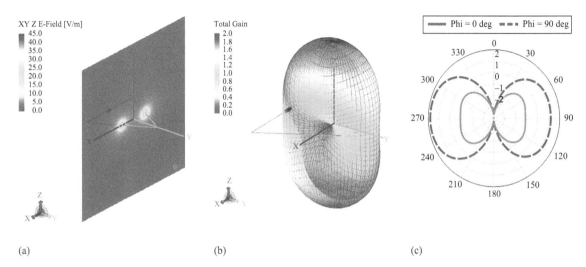

FIGURE 3-23 ▪ Radiation performance of a center-fed triangular loop antenna with $C = \lambda$ in POSTFEKO: (a) Near-field distribution in the y-z plane. (b) 3D gain pattern. (c) Gain radiation patterns in x-z and y-z planes.

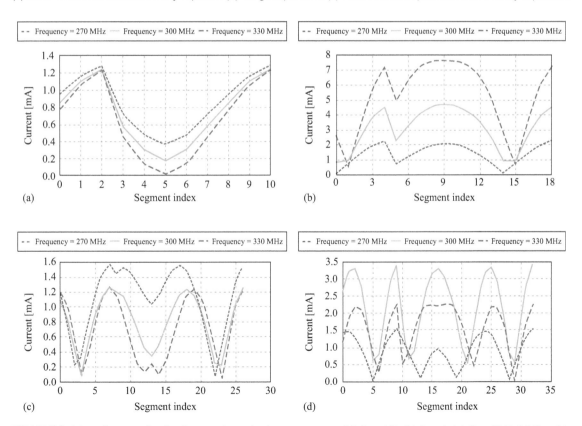

FIGURE 3-24 ▪ Current distribution on triangular loop antennas: (a) $C = \lambda/2$. (b) $C = \lambda$. (c) $C = 3\lambda/2$. (d) $C = 2\lambda$.

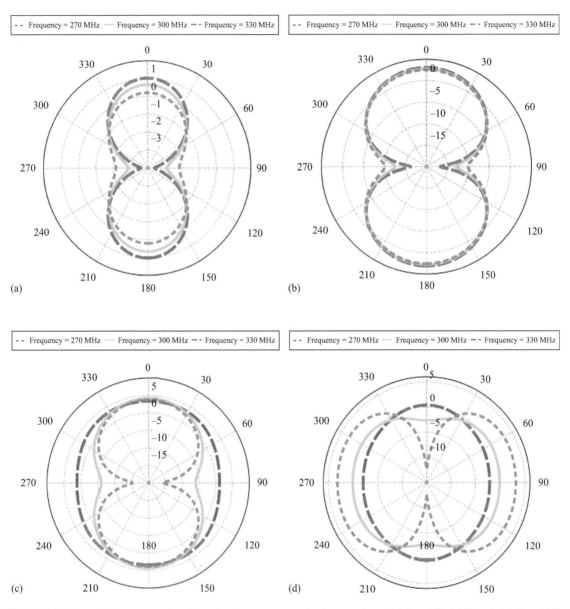

FIGURE 3-25 ▪ Gain patterns for triangular loop antennas in the x-z plane: (a) $C = \lambda/2$. (b) $C = \lambda$. (c) $C = 3\lambda/2$. (d) $C = 2\lambda$.

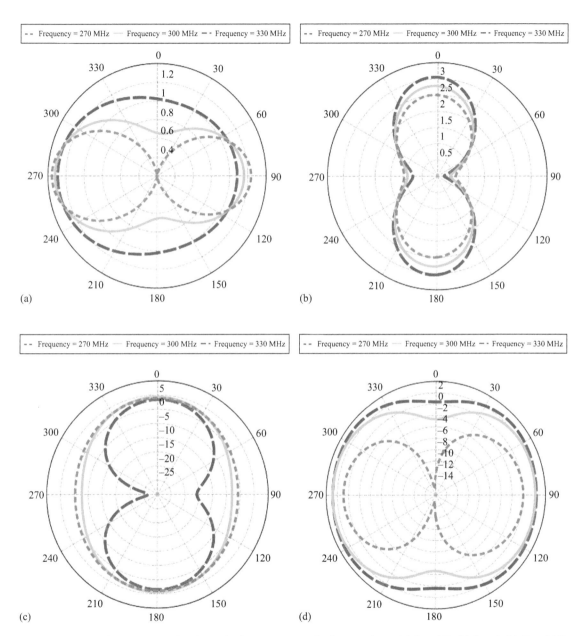

FIGURE 3-26 ■ Gain patterns for triangular loop antennas in the y-z plane: (a) $C = \lambda/2$. (b) $C = \lambda$. (c) $C = 3\lambda/2$. (d) $C = 2\lambda$.

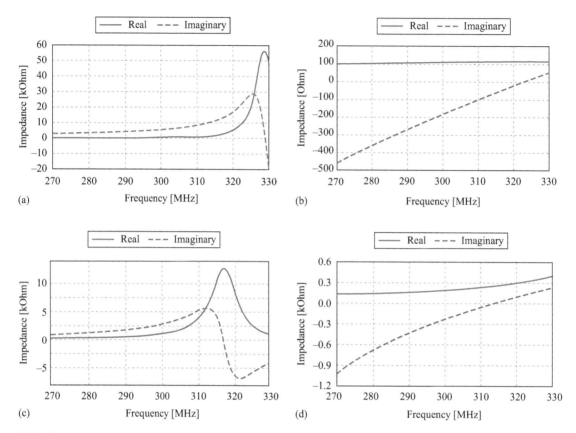

FIGURE 3-27 ■ Impedance as a function of frequency for triangular loop antennas: (a) $C = \lambda/2$. (b) $C = \lambda$. (c) $C = 3\lambda/2$. (d) $C = 2\lambda$.

3.6 | LOOP ANTENNAS NEAR A PEC SCATTERER

As discussed in Chapter 2, in many practical applications the antenna is placed in a medium where surrounding objects will distort the radiation pattern of the antenna. Here we will study the radiation performance of a UHF circular loop antenna with $C = \lambda$ when it is placed near PEC scatterers. The geometrical model of the problem is shown in Figure 3-28.

First let's study a finite-length PEC cylinder with a radius of 0.5λ and length of 2λ, which is placed at a distance (λ) from the center of the loop and the coordinate system; that is, the minimum distance between the cylinder and the loop is 0.5λ. The current distribution on the cylinder and the radiation pattern is shown in Figure 3-29. As discussed before, due to the presence of this large

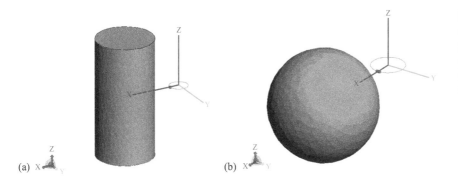

FIGURE 3-28 ▤
A loop antenna placed near a PEC scatterer: (a) Finite-length cylinder. (b) Sphere.

(a)

(b)

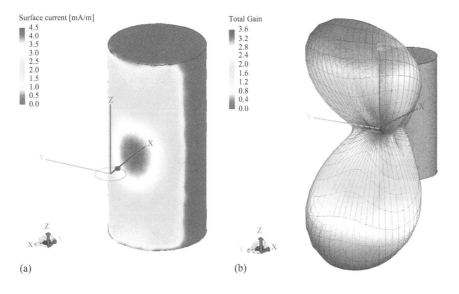

FIGURE 3-29 ▤
(a) Current induced on the surface of a PEC cylinder due to radiation from a loop antenna. (b) Radiation pattern of a loop antenna when placed close to a PEC cylinder.

(a)

(b)

scattering object the radiation pattern of the loop antenna will be distorted. However, compared with the dipole case, the peak gain of the antenna has not increased much.

Next we study the case when a PEC sphere with a radius of 0.5λ is placed at a distance (λ) from the center of the loop and the coordinate system. The current distribution on the sphere and the radiation pattern is shown in Figure 3-30. Again, compared with the dipole, in this case the increase in gain is not significant. In addition, comparing the radiation patterns for these two cases reveals a noticeable difference when the scatterer object is changed; as a result, in general loop antennas are not suitable feeds for reflector antennas.

It is also interesting to study the radiation performance of the loop when it is placed near a planar scatterer. Here we consider a rectangular PEC sheet with a

FIGURE 3-30 ▪
(a) Current induced on the surface of a PEC sphere due to radiation from a loop antenna.
(b) Radiation pattern of a loop antenna when placed close to a PEC sphere.

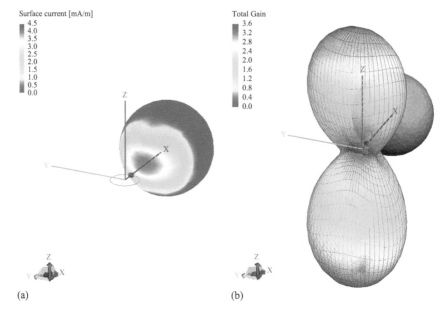

(a)   (b)

FIGURE 3-31 ▪
(a) Current induced on the surface of a PEC sheet due to radiation from a loop antenna.
(b) Radiation pattern of a loop antenna when placed close to a PEC sheet.

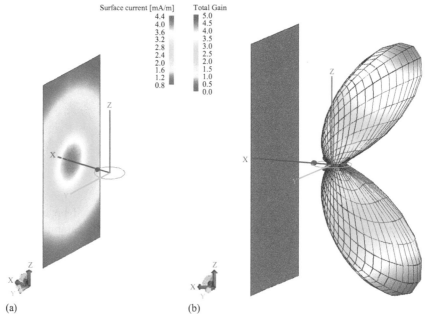

(a)   (b)

width of λ and a length of 2λ. The scatterer is placed at a distance (λ) from the center of the loop and the coordinate system. The current distribution on the sheet and the radiation pattern is provided in Figure 3-31. Comparison of these results with the cylinder case (Figure 3-29) highlight that the blockage aperture size is very similar (but not identical) but the radiation pattern is quite different. Moreover, with this configuration a slightly higher gain can be achieved.

EXERCISES

(1) *Effects of Ground Plane.* For 300 MHz operation, design a one-wavelength circular loop antenna and place it such that the plane containing the loop is parallel to the $z = 0$ plane at a distance of $\lambda/2$ from the plane. Use the infinite ground plane model (x-y plane) and calculate the radiation characteristics of the loop antenna. What is the difference in the antenna performance compared with the loop placed in free space? Repeat this exercise when the plane containing the loop is normal to the ground plane.

(2) *Small Loop Antennas.* For 900 MHz operation, design three loop antennas, with circular, square, and triangular shapes, all with a circumference of $\lambda/10$. For the square and triangular loops use the center-fed configuration. Compare the radiation patterns of all three loops at center and off-center frequencies. Are the radiation patterns of these three different configurations similar? Explain your answer.

(3) *Triangular Loop Antenna.* For the triangular loop antenna studied in section 3.5, place the port at the corner of the triangle and compare the performance of this loop with the center-fed case. Which design achieves a better performance? Explain your answer.

(4) *Loop Antennas Placed Near a PEC Scatterer.* Consider the same scattering objects studied in section 3.6, but place the plane of the loop antenna normal to the x-y plane. Compare this with the results presented in section 3.6 and discuss your observations.

(5) *Wireless Power Transfer.* Repeat the study performed in section 2.5, with two circular loop antennas each having 1λ circumference. What should the orientation of the loop antennas be (with respect to each other) to achieve maximum power transfer? Compared with the dipole case, does the received power at the port increase? Explain your answer.

Microstrip Patch Antennas

Chapter Outline

4.1 | INTRODUCTION

With the rapid development of printed circuit technology, printed antennas have found numerous applications in high-performance systems where size, weight, and cost are of significant importance [13–17]. The most notable feature of these printed antennas is that they are low profile. In particular, microstrip patch antennas and arrays of patches are perhaps the most common form of these printed structures. Extensive investigations into the radiation characteristics of patch antennas began in the 1970s, and both analytical and numerical solutions have been developed for printed antennas [7,8]. Moreover, the physical insight obtained by these analyses led to the development of several advanced design configurations. While the conventional microstrip patch antenna has a narrow bandwidth, their advanced configurations can achieve broadband and multiband performance. In addition, both circular and linear polarization can be achieved with the patch antenna. In this chapter, we will briefly study some of the fundamentals of microstrip patch antennas, will show the procedure for designing patch antennas using the commercial software package FEKO [9], and will present several examples.

4.2 | PATCH ANTENNA DESIGN AND ANALYSIS

The basic configuration of a patch antenna consists of a very thin metallic strip (patch) that is placed a small fraction of a wavelength above a metallic ground plane. The patch and the ground plane are typically separated by a dielectric substrate. The conventional rectangular patch antenna has a length less than $\lambda/2$. A geometrical model of a patch antenna is given in Figure 4-1.

FIGURE 4-1 ■ Geometrical arrangement of a rectangular microstrip patch antenna.

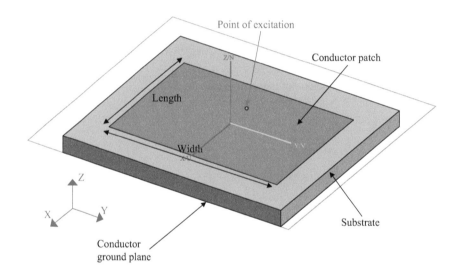

Various designs can be used to feed a microstrip patch antenna, the most common of which are the coaxial probe, microstrip line, aperture coupling, and proximity coupling [7]. For an accurate analysis of the microstrip antenna, it is imperative that the feed be modeled accurately. While circuit models are available for these four feed configurations, a more accurate analysis would generally require a full-wave simulation.

Different analytical methods have been developed over the years for microstrip antennas, the most popular of which are transmission-line, cavity, and full-wave. The transmission-line method is the simplest of all and gives good physical insight, but it is generally the least accurate. The cavity model is more accurate but at the same time more complex. Nonetheless, it also gives a good physical insight. The analytical formulations for the transmission-line and cavity model analysis of patch antenna radiation characteristics are available in the literature [7,8], and readers are encouraged to become familiar with these methods for better understanding. This chapter, however, is mainly focused on microstrip patch antenna full-wave simulation using the commercial software FEKO, but we will also outline the basic guidelines and equations for designing a rectangular microstrip antenna.

The first step is to select a suitable substrate for the design. Commercial laminates are available in a variety of thicknesses with different electrical properties. Generally, it is preferable to use dielectric materials with a lower dielectric constant since this will provide a wider impedance bandwidth and will reduce surface wave excitation. High dielectric constant materials are typically used when the physical restraints necessitates reducing the antenna size. Once the substrate material is selected, the next task is to choose a suitable thickness for the laminate. In microstrip antennas, thicker substrates usually provide a wider bandwidth and are preferable, but matching is an issue for very thick substrates. A general rule of thumb for coaxial fed microstrip antennas is to select a thickness less than $0.03\lambda_0$, which is considered to be a thin substrate.

With these design properties specified, the task is to determine the width and length of the patch. For good radiation efficiency, the practical width of the patch is approximately determined using

$$W = \frac{1}{2f_r\sqrt{\mu_0\varepsilon_0}}\sqrt{\frac{2}{\varepsilon_r+1}}, \qquad (4\text{-}1)$$

where f_r is the resonant frequency of the patch antenna, ε_0 and μ_0 are the permittivity and permeability in free space, respectively, and ε_r is the dielectric constant of the substrate. In general, there is some flexibility in selecting the patch width, and depending on the design even square geometry may be used.

In the next stage we need to determine the resonant length of the patch. An important consideration in designing a microstrip antenna is to accurately take into account the effect of the fringing fields, which generate an electrically wider dimension for the patch. Since some of the waves travel in the substrate and some in air, an effective dielectric constant is introduced to account for fringing and the wave propagation in the line. The effective dielectric constant can be computed using

$$\varepsilon_{reff} = \frac{\varepsilon_r+1}{2} + \frac{\varepsilon_r-1}{2}\left[1+12\frac{h}{W}\right]^{-1/2}, \qquad (4\text{-}2)$$

where h is the thickness of the substrate. The electrical length increase of the microstrip patch can then be computed using the empirical formula

$$\frac{\Delta L}{h} = 0.412 \frac{(\varepsilon_{reff} + 0.3)\left(\dfrac{W}{h} + 0.264\right)}{(\varepsilon_{reff} - 0.258)\left(\dfrac{W}{h} + 0.8\right)}. \qquad (4\text{-}3)$$

The fringing field effect results in an increase of ΔL on both sides of the patch length; therefore, the effective length of the patch will be

$$L_{eff} = L + 2\Delta L. \qquad (4\text{-}4)$$

In summary, the design procedure for a rectangular microstrip patch is as follows:

(1) Determine the required width using equation (4-1).
(2) Calculate the effective dielectric constant using equation (4-2).
(3) Compute the electrical increase in patch length using equation (4-3).
(4) Calculate the physical length of the patch using

$$L = L_{eff} - 2\Delta L, \qquad (4\text{-}5)$$

where the effective length of the patch is given by

$$L_{eff} = \frac{1}{2 f_r \sqrt{\varepsilon_{reff}} \sqrt{\mu_0 \varepsilon_0}}. \qquad (4\text{-}6)$$

As a concluding remark in this section, it should be noted that the feed position is determined by the impedance characteristics of the microstrip patch. That is because the variation of input resistance at resonance with feed position follows that of the cavity field. For the lowest mode, which is typically used, the input resistance is maximum at the edge of the patch and decreases as the feed moves inside the patch.

4.3 | FULL-WAVE SIMULATION OF PATCH ANTENNAS IN FEKO

4.3.1 A Rectangular Microstrip Patch Antenna with Probe-Fed Excitation

This section outlines the design and simulation of a rectangular microstrip patch antenna for the operating frequency of 2.45 GHz. For the dielectric substrate we select a laminate with a dielectric constant of 2.2 and a thickness of 1.115 mm. The latter corresponds to an electrical thickness of about $0.01\lambda_0$ and therefore should provide a good matching at the design frequency. The size of the substrate and ground plane is selected to be 60.0×70.0 mm. The initial dimensions of the patch are 40.76×48.40 mm, which were determined using the procedure outlined in the previous section.

The patch antenna can be modeled quite simply in FEKO by creating a cuboid for the substrate and a rectangle for the patch. One important consideration, however, is the probe feed. The probe is modeled by using a wire that connects the ground conductor (under the substrate) to the bottom surface of the patch. Once these three geometries are created, they must be united to ensure proper connection of the mesh. A wire port and a voltage source are then defined to excite the antenna. The geometrical model of this rectangular probe-fed patch antenna is shown in Figure 4-2.

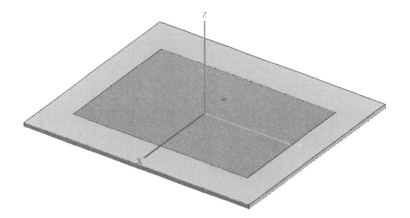

FIGURE 4-2 ▪
A probe-fed rectangular microstrip patch antenna in FEKO.

As discussed earlier, the initial dimensions are merely a starting point for the design, and further tuning is necessary. After some parametric studies, the dimensions of the patch were set to 38.9×53.95 mm. Moreover, the feed position was also tuned to provide optimum matching. The final position of the feed is at $x = -7.25$ mm, based on the coordinate system shown in Figure 4-2. The input impedance and reflection coefficient are given in Figure 4-3, which illustrates that at the resonant frequency the imaginary part of the impedance is zero and the real part is maximum. Also, the sharp variation of impedance indicates that the bandwidth is quite narrow (Figure 4-3b).

It is also interesting to see the currents on the patch at the resonance frequency (Figure 4-4). Along the length of the antenna, a maximum current at the center of the edge corresponds to the fundamental (lowest) mode excitation.

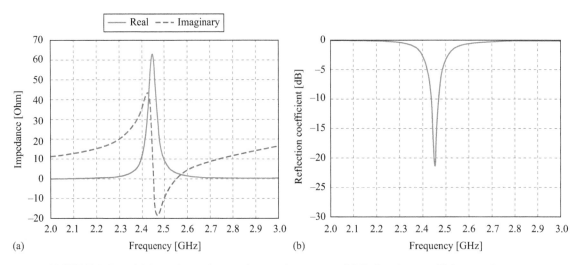

(a)

(b)

FIGURE 4-3 ▪ (a) Input impedance of a patch antenna. (b) Reflection coefficient at the port.

FIGURE 4-4 ▪
Currents on a
rectangular patch
at resonance.

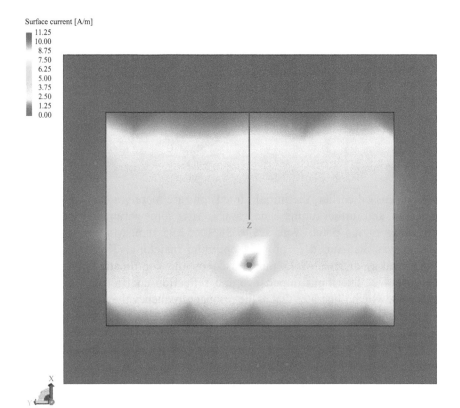

The radiation pattern of the antenna is shown in Figure 4-5. The antenna's radiation pattern is in the broadside direction (normal to the aperture of the patch). In addition, the pattern is quite symmetric in both principal planes, which is a very practical feature of patch antennas. The antenna's 3D radiation pattern is illustrated in Figure 4-6. The antenna gain as a function of frequency is also given in Figure 4-7. The maximum realized gain is 7.35 dBi, and the 3 dB gain bandwidth of this antenna is about 100 MHz. The spike observed in Figure 4-7 at 2.25 GHz is due to the numerical errors when using the frequency interpolation option. A discrete frequency sweep may be used to correct this problem.

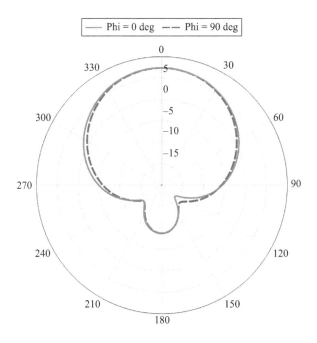

FIGURE 4-5 ▪ Radiation patterns of a rectangular patch antenna in the principal planes at 2.45 GHz.

4.3.2 A Rectangular Microstrip Patch Antenna with Microstrip-Line Excitation

Here we will study a different feed configuration for the microstrip patch antenna: the microstrip-line excitation. Like the probe feed, the microstrip feed is easy to fabricate and match. Matching is achieved using a recessed (i.e., inset) feed, as will be demonstrated here. To be consistent with the previous example, we design the antenna for the same operating frequency of 2.45 GHz and employ the same substrate material. Figure 4-8 shows the geometrical

FIGURE 4-6 ▪ 3D radiation pattern of a rectangular patch antenna at 2.45 GHz.

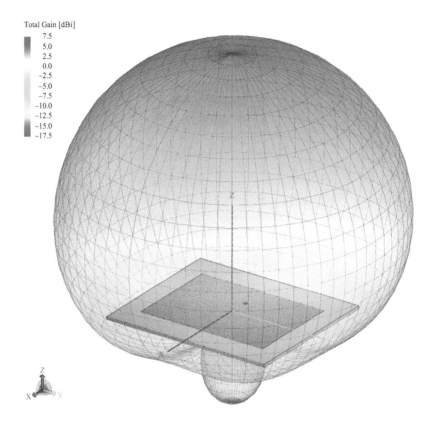

FIGURE 4-7 ▪ Gain of a rectangular microstrip patch antenna as a function of frequency.

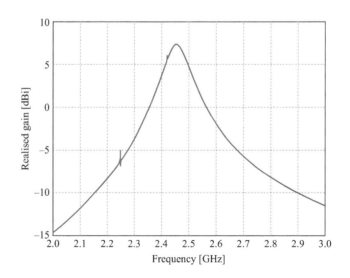

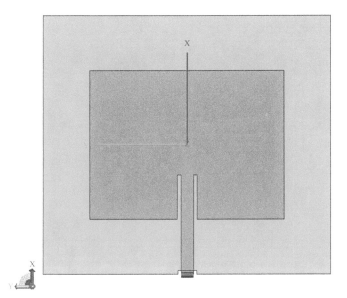

FIGURE 4-8 ▪
A microstrip feed for a rectangular patch antenna configuration in FEKO.

model of this antenna as well as how the microstrip feed is moved into the patch to provide the required impedance matching. Since the feed method is different, it is necessary to slightly tune the antenna length. After some parametric studies, the length of the patch was set to 40.2 mm. However, the patch width was not changed ($W = 53.95$ mm). The length of the inset section is 12 mm, which corresponds to an offset of 8.1 mm. This is quite close to the value of 7.25 mm used for the probe-fed design. The gap width for this inset is 1 mm on both sides.

Modeling this antenna in FEKO is straightforward, but some attention must be given to the feed section. To ensure proper excitation, the microstrip line has to be excited with an edge port. However, it cannot be defined on the dielectric boundary of the substrate (please refer to the FEKO manual for more details on this issue), so a small gap is cut out of the substrate. This is shown in Figure 4-9.

The input impedance and reflection coefficient are shown in Figure 4-10, where, similar to the previous case, at the resonant frequency the imaginary part of the impedance is zero and the real part is maximum. Again, the sharp variation of impedance is the reason for the narrow bandwidth observed in Figure 4-10b.

The electric currents on the patch at the resonance frequency are shown in Figure 4-11. Similar to the probe-fed design, a maximum current is observed at the center of both edges, which is indicative of the excitation of the lower mode.

The 2D and 3D radiation patterns of the antenna are shown in Figure 4-12 and Figure 4-13, respectively. A broadside, almost symmetric radiation pattern is obtained for this antenna.

FIGURE 4-9 ▪
Model of a
microstrip feed
excitation in FEKO.

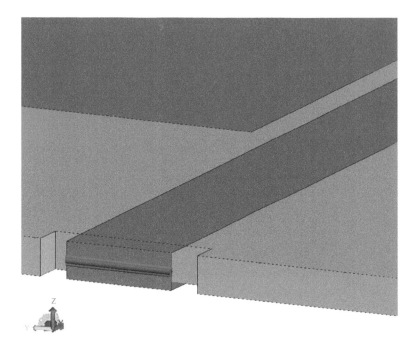

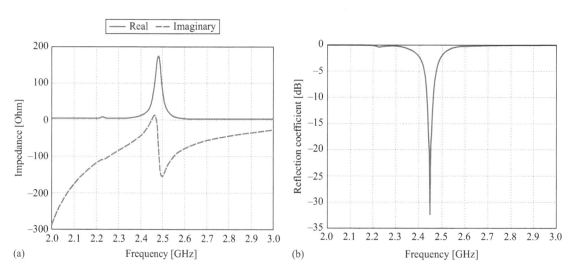

(a)

(b)

FIGURE 4-10 ▪ (a) Input impedance of a patch antenna. (b) Reflection coefficient at the port.

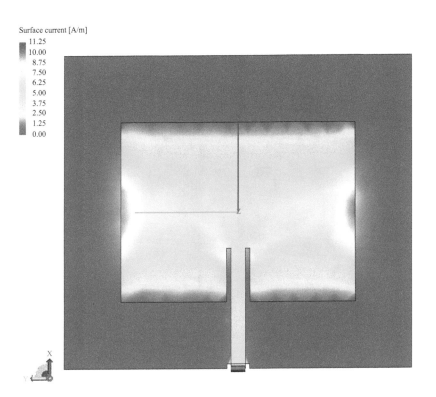

Surface current [A/m]

FIGURE 4-11 ▪
Currents on a
rectangular patch
at resonance.

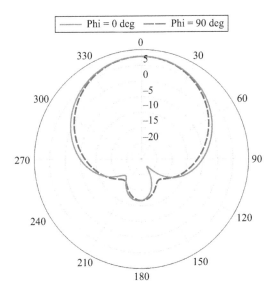

FIGURE 4-12 ▪
Radiation patterns of
a rectangular patch
antenna in the
principal planes at
2.45 GHz.

FIGURE 4-13 ▪ 3D radiation pattern of a rectangular patch antenna at 2.45 GHz.

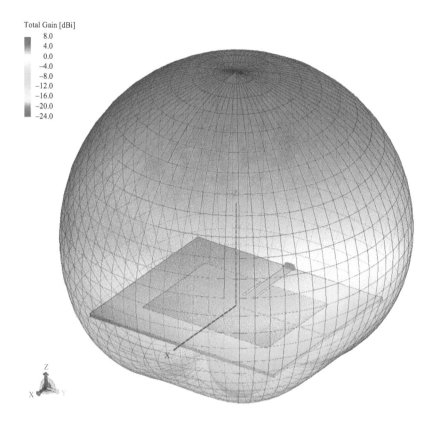

Total Gain [dBi]

8.0
4.0
0.0
−4.0
−8.0
−12.0
−16.0
−20.0
−24.0

The antenna gain as a function of frequency is provided in Figure 4-14, where the maximum realized gain is 7.83 dBi and the 3 dB gain bandwidth of this antenna is about 80 MHz. The realized gain of the antenna is the gain taking into account the reflection losses at the input of the antenna.

Importantly, while the microstrip-fed design presented here doesn't have much advantage over the probe-fed design as a single antenna element, the great advantage of this feed method is in microstrip array configurations. Similar to the discussion earlier, the spikes observed in Figure 4-14 are due to the numerical errors when using the frequency interpolation, which can be fixed using a discrete frequency sweep.

4.3.3 A Circular Microstrip Patch Antenna with Probe-Fed Excitation

Another popular configuration for a microstrip antenna is a circular patch or disc. The analysis and design equations are based on the cavity model as described in [7], and readers are encouraged to study the cited reference. Here we will study the performance of a circular microstrip patch designed for the operating

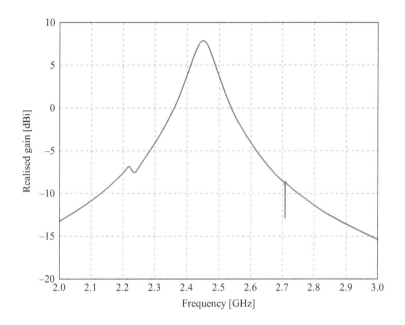

FIGURE 4-14 ▪
Gain of a rectangular
microstrip patch
antenna as a
function of
frequency.

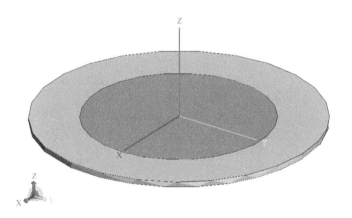

FIGURE 4-15 ▪
A probe-fed circular
microstrip patch
antenna in FEKO.

frequency of 2.45 GHz. We select the same dielectric material as the previous examples, but with a slightly thicker substrate (i.e., 1.26 mm). The geometrical model of this circular probe-fed patch antenna is shown in Figure 4-15.

The radius of the patch and the circular ground plane are 23 mm and 35 mm, respectively. The probe feed is placed at a distance of $x = -10$ mm from the center of the antenna patch. The input impedance and reflection coefficient are shown in Figure 4-16, where a similar observation as the previous cases can be made.

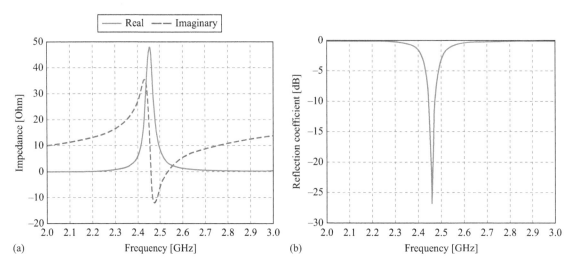

(a) Frequency [GHz]

(b) Frequency [GHz]

FIGURE 4-16 ▪ (a) Input impedance of a patch antenna. (b) Reflection coefficient at the port.

FIGURE 4-17 ▪
Currents on a
circular patch at
resonance.

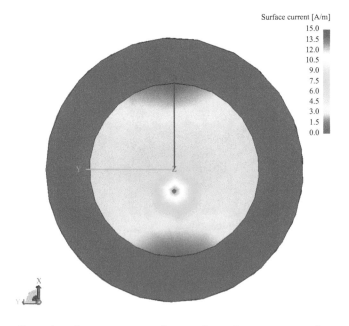

The surface electric currents on the patch at the resonance frequency are
shown in Figure 4-17, which corresponds to the lower mode excitation for the
circular patch.

The 2D and 3D radiation patterns of the antenna are shown in Figure 4-18
and Figure 4-19, respectively. Similar to the rectangular patch configurations,
a broadside, almost symmetric radiation pattern is also obtained with the circular patch antenna.

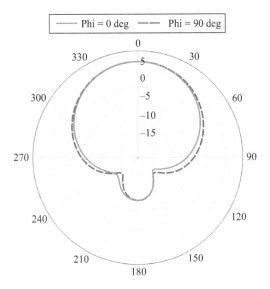

FIGURE 4-18 ▪
Radiation pattern of
a circular patch
antenna in the
principal planes at
2.45 GHz.

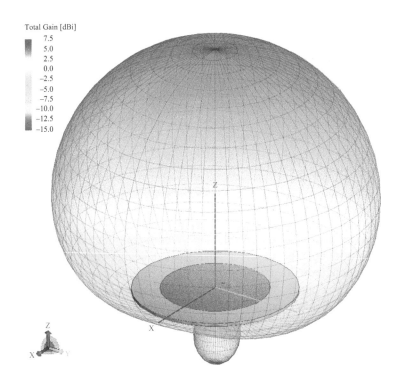

FIGURE 4-19 ▪
3D radiation pattern
of a circular patch
antenna at 2.45 GHz.

The antenna gain as a function of frequency is also given in Figure 4-20, where the maximum realized gain is 7.04 dB and the 3 dB gain bandwidth of this antenna is about 82 MHz.

FIGURE 4-20 ▪
Gain of a circular microstrip patch antenna as a function of frequency.

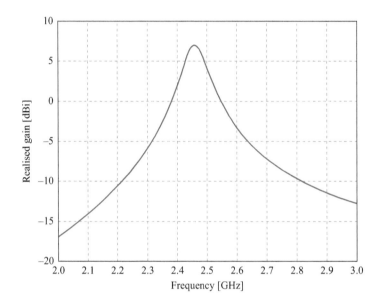

▮ 4.4 | CIRCULARLY POLARIZED PATCH ANTENNAS

Circularly polarized (CP) antennas are desirable in certain applications to reduce the polarization loss caused by the misalignment between the transmitting and the receiving antennas. In terms of feed methods for circularly polarized patch antennas, single- and dual-feed configurations each have their own advantages [7,8]. Here we will demonstrate the performance of a square patch antenna with dual-fed configuration, which can realize circular polarization.

The geometrical model of this dual-feed patch antenna is shown in Figure 4-21. In contrast to the microstrip-fed design demonstrated in the previous section, here the transmission lines are directly connected to the edge of the microstrip patch. With such a feed configuration, it is necessary to use quarter wavelength transmission lines to provide proper matching to the impedance of the ports.

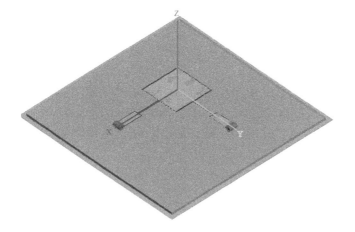

FIGURE 4-21 ▣
A dual-feed square
patch antenna
modeled in FEKO.

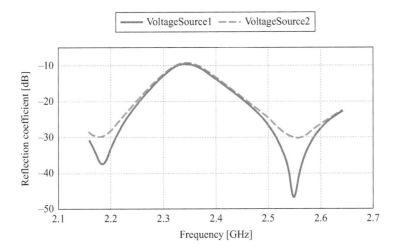

FIGURE 4-22 ▣
Reflection coefficient
at the two ports.

The center design frequency is 2.4 GHz, and a substrate with dielectric constant of 2 and a thickness of 2.6498 mm is selected for this design. The first section of the transmission lines (which is connected to the 50 Ω ports) has a width of 8.67 mm, which corresponds to a characteristic impedance of about 50 Ω. The width of the quarter-wavelength section is 1.56 mm. The patch antenna has a square aperture with a side length of 42.55 mm. An infinite substrate model is used for this design. Figure 4-22 illustrates the simulated reflection coefficients at the two ports; the antenna is matched quite well across the band.

The electric currents on the patch at 2.37 GHz are given in Figure 4-23.

FIGURE 4-23 ▪
Electric currents on a dual-fed square patch antenna.

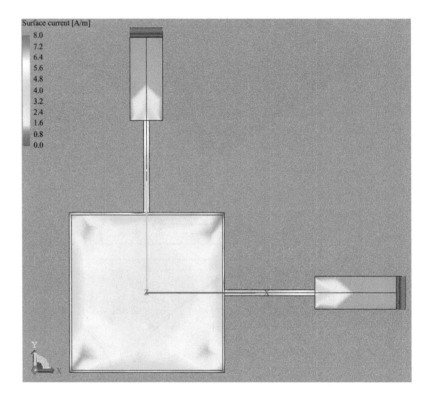

For this two-feed configuration, circular polarization can easily be achieved by providing a 90° phase difference at the two ports. Here a 90° phase delay is set for the port along the y axis, which results in a right-hand circularly polarized radiation pattern for the antenna. The axial ratio of the antenna as a function of frequency is shown in Figure 4-24. Very good circularly polarized performance is obtained with this design. Figure 4-25 outlines the right-hand circularly polarized gain.

The radiation pattern of the antenna at 2.37 GHz is presented in Figure 4-26. Note that the cross-polarization is about 30 dB down in the broadside direction and increases as it approaches the horizontal plane. Nonetheless, this circularly polarized antenna provides good hemispherical coverage, which makes it quite suitable for many applications. The 3D radiation pattern of this antenna is shown in Figure 4-27.

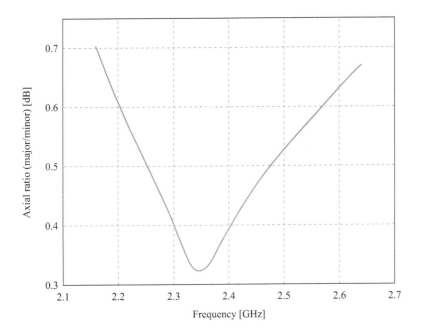

FIGURE 4-24 ▪
Axial ratio of a CP
patch antenna as a
function of
frequency.

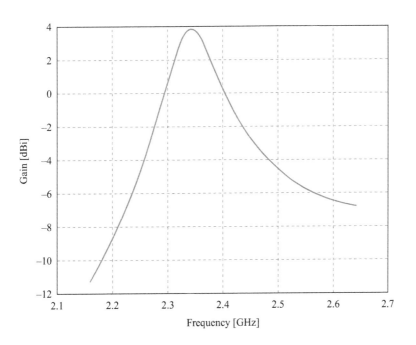

FIGURE 4-25 ▪
Right-hand circularly
polarized gain as a
function of
frequency.

FIGURE 4-26 ▪
Radiation patterns in
the principal planes.

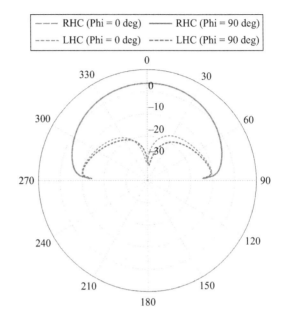

FIGURE 4-27 ▪ 3D
radiation pattern of a
CP patch antenna.

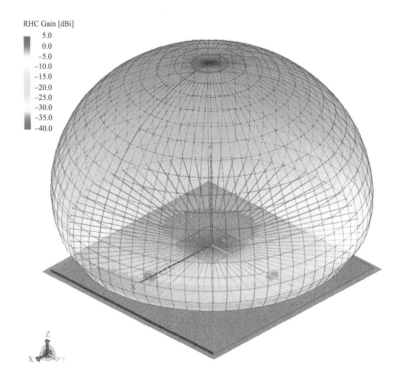

EXERCISES

(1) ***Rectangular Patch Antennas.*** Following the design example given in section 4.3, design a probe-fed rectangular patch antenna for the operating frequency of 5.8 GHz. Select a commercially available laminate and study the effect of different substrate thickness for the design. Does using thicker substrate improve the bandwidth of the antenna? Explain your answer.

(2) ***Broadband Patch Antennas.*** An inherent limitation of microstrip antennas is the narrow impedance bandwidth, which is typically only a couple of percent. One design that can improve the bandwidth of a microstrip patch antenna is the U-slot patch configuration. The geometrical model of this antenna is shown in Figure P4-1. Study the performance of the U-slot antenna based on the dimensions given in [15] and verify that a broadband performance can be achieved with this design.

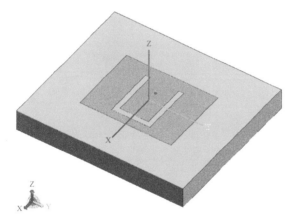

FIGURE P4-1 ▪ A U-slot patch antenna.

(3) ***Circularly Polarized Patch Antennas.*** Follow the design example in section 4.4, and design a dual-fed circularly polarized patch antenna for the operating frequency of 5.8 GHz. Use the recessed microstrip feed configuration presented in section 4.3 for the design.

Microstrip-Based Feeding Networks

Chapter Outline

5.1 | INTRODUCTION

For many antenna configurations (e.g., antenna arrays or multiport antennas) it is necessary to have a mechanism to divide the input power between the antenna ports. This is generally referred to as the feed network. Power dividers are passive microwave components that are used for power division (transmit) or power combining (receive) in array antennas. In this chapter we will briefly review the basics of microstrip transmission lines and then will study several different power dividers that are practical for the feed networks of microstrip array antennas.

5.2 | DESIGN OF MICROSTRIP TRANSMISSION LINES

The microstrip line is perhaps the most popular type of planar transmission line. In particular, it can directly feed microstrip patch antennas in an integrated fabrication process. By comparison, patch antenna probe feeding typically complicates fabrications, particularly at higher frequencies. In addition, the microstrip line can be easily integrated with passive or active microwave devices [18]. The basic geometry of a microstrip transmission line is shown in Figure 5-1. It consists of a grounded dielectric substrate layer, a conductor of width (W), and a length that could be infinite in theory.

FIGURE 5-1 ■
Geometrical model of a microstrip transmission line.

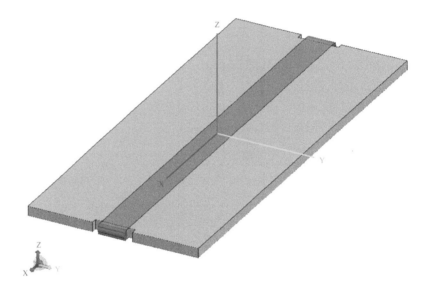

Most of the fields of microstrip transmission lines with a thin substrate are generally concentrated between the strip and the ground plane; as such, a quasi-transverse electromagnetic (TEM) approximation is typically acceptable. The exact fields of a microstrip line, however, are hybrid and require advanced analysis techniques. On the other hand, because of the practicality of the microstrip transmission lines, empirical formulas have also been derived, and they are outlined here. The primary concern when designing a microstrip line is to obtain the characteristic impedance of the transmission line. In addition, it is typically necessary to know the electrical length of the line, which requires calculating the effective dielectric constant.

The effective dielectric constant of a microstrip line is approximately given by [18]

$$\varepsilon_e = \frac{\varepsilon_r + 1}{2} + \frac{\varepsilon_r - 1}{2}\left[1 + 12\frac{h}{W}\right]^{-1/2}, \qquad (5\text{-}1)$$

where h is the substrate thickness, W is the width of the microstrip, and ε_r is the dielectric constant of the substrate. This is identical to equation (4-2) for the microstrip patch antenna. The characteristic impedance (Z_0) of the microstrip transmission line is then approximately obtained using [18]

$$Z_0 = \begin{cases} \dfrac{60}{\sqrt{\varepsilon_e}} \ln\left(\dfrac{8h}{W} + \dfrac{W}{4h}\right) & \text{for } \dfrac{W}{h} \le 1 \\[3ex] \dfrac{120\pi}{\sqrt{\varepsilon_e}[W/h + 1.393 + 0.667\ln(W/h + 1.444)]} & \text{for } \dfrac{W}{h} \ge 1 \end{cases}. \quad (5\text{-}2)$$

The electrical length of the line is βl, where β is the propagation constant, and l is the physical length of the transmission line. To achieve a phase shift of ϕ, the length of the line can be computed using

$$l = \frac{\phi}{\beta} = \frac{\phi}{\sqrt{\varepsilon_e}\dfrac{2\pi f}{c}}, \quad (5\text{-}3)$$

where f is the design frequency, and c is the speed of light in free space.

As an example, we will consider a design for a microstrip transmission line with a characteristic impedance of 50 Ω for 2.45 GHz operation using a substrate with a dielectric constant of 2.2 and a thickness of 1.5748 mm. For this design, the width of the microstrip is 4.852 mm. The geometrical model of this microstrip transmission line in FEKO is given in Figure 5-2. In practice, the transmission lines would have a finite length. Two edge ports are defined for this transmission line, as described in Chapter 4. The voltage source with an internal impedance of 50 Ω is added to the first port (left port in Figure 5-2), and the second port is terminated with a 50 Ω load. In general, this setup would mimic an infinite transmission line; that is, no reflections are expected from the ports.

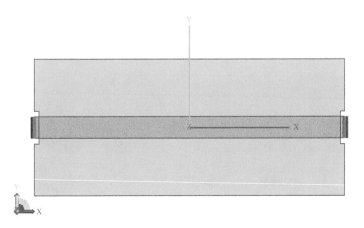

FIGURE 5-2 ▪
A 50 Ω microstrip transmission line.

The characteristic impedance of the line can be observed by looking at the input impedance at the voltage source. This is shown in Figure 5-3 across the frequency band. The real part of the impedance is almost 50 Ω at the center design frequency and slightly changes across the entire frequency band.

FIGURE 5-3 ▪
Impedance of a
microstrip
transmission line as
a function of
frequency.

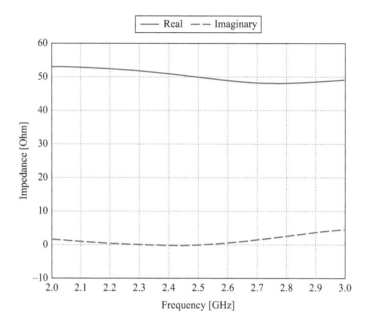

The imaginary part is also zero at the center frequency and slightly varies
around the zero value across the entire frequency band. Consequently, this
would mean that good matching and transmission can be achieved with this
design. The reflection and transmission coefficients are given in Figure 5-4,
which shows that a very good transmission (0 dB) and a good matching (about
−30 dB) are achieved across the band.

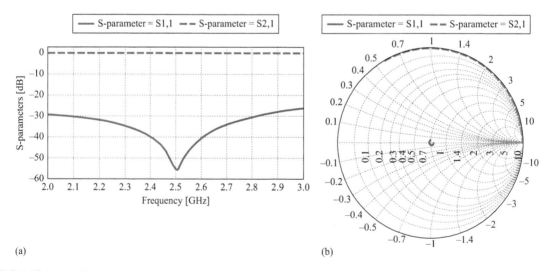

(a)

(b)

FIGURE 5-4 ▪ Reflection and transmission coefficients of the microstrip transmission line as a function of
frequency: (a) Cartesian plot. (b) Smith chart.

The electric current distribution on the top surface of the substrate at 2.45 GHz is shown in Figure 5-5. As discussed earlier, the fields are almost completely confined beneath the microstrip transmission line.

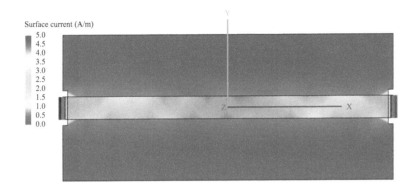

Surface current (A/m)

FIGURE 5-5 ▪
Electric currents at 2.45 GHz.

5.3 | THE QUARTER-WAVELENGTH TRANSFORMER

The quarter-wavelength transformer is a very practical circuit for impedance matching. In many microwave applications it is necessary to match a load (R_L) to a transmission line with certain characteristic impedance, say, Z_0. A quarter-wavelength transformer is a transmission line with a length of $\lambda/4$ (or phase of 90°) and impedance equal to

$$Z_{QW} = \sqrt{Z_0 R_L}. \qquad (5\text{-}4)$$

Such a transmission line can provide a perfect matching at the center design frequency. As an example of a quarter-wavelength transformer, we consider a design matching a 50 Ω transmission line to a 100 Ω load at 2.45 GHz. The geometrical model of this is shown in Figure 5-6. For this design we use the same substrate material and thickness as in section 5.2. The quarter-wavelength microstrip transformer should have a characteristic impedance of 70.71 Ω, which corresponds to a width of 2.783 mm. The length of the line should also be 22.722 mm, corresponding to an electrical length of quarter-wavelength. The transformer is connected on one side to a 50 Ω transmission line excited by a voltage source port. The length of the transmission line is 23.639 mm. On the other side the transformer is connected to a 100 Ω transmission line with a width of 1.411 mm, is terminated with a port with a 100 Ω load, and has a transmission line of length 23.639 mm. In total, the length of this microstrip circuit is 70 mm. The width of the substrate is

FIGURE 5-6 ▪
A quarter-
wavelength
microstrip
transformer.

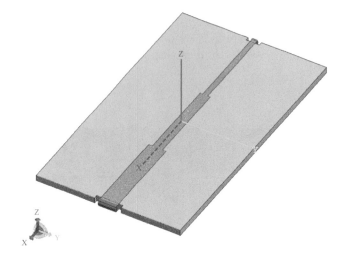

FIGURE 5-6 ▪
A quarter-
wavelength
microstrip
transformer.

FIGURE 5-7 ▪
Reflection and
transmission
coefficients of a
microstrip quarter-
wavelength
transformer.

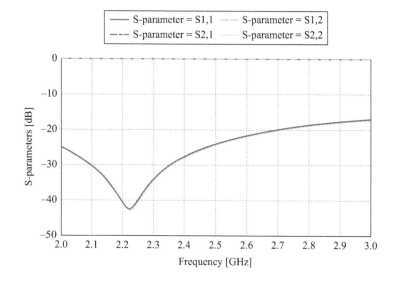

selected to be 40 mm, but in general it is possible to use a smaller width if
required.

The reflection and transmission coefficients for this two-port design are
shown in Figure 5-7. At the center frequency, a very good reflection coefficient
is obtained for the design. Typically, however, this matching technique is narrowband. The electric current distribution on the transmission lines is also
given in Figure 5-8.

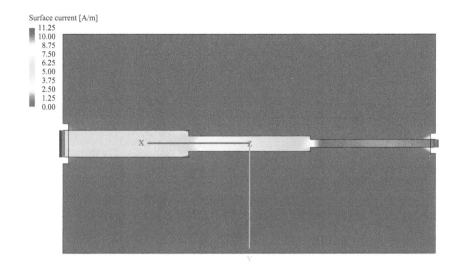

Surface current [A/m]
11.25
10.00
8.75
7.50
6.25
5.00
3.75
2.50
1.25
0.00

FIGURE 5-8 ▪
Electric currents on a quarter-wavelength microstrip transmission line.

5.4 | T-JUNCTION POWER DIVIDERS

Another very practical circuit for antenna feed networks is the T-junction power divider. This configuration is a three-port microwave network that can be used for power division or power combining. For lossless (or low-loss) transmission lines that have real values of characteristic impedances, and ignoring the stored energy at the line discontinuities, the equation for determining the impedance of each port is

$$\frac{1}{Z_1} + \frac{1}{Z_2} = \frac{1}{Z_0}. \tag{5-5}$$

Consider a power divider where Z_0 is the characteristic impedance of the input line and Z_1 and Z_2 are the characteristic impedances of the output lines. By selecting the output line impedances, the input power can be split as desired.

As an example, consider a microstrip T-junction divider designed to achieve an equal power split (i.e., each out port supports −3 dB relative to the input). We will use the same substrate material and thickness as in section 5.2. The input line has a characteristic impedance of 50 Ω; thus, both output lines need to have a 100 Ω impedance to satisfy equation (5-5). This corresponds to strip widths of 4.743 mm and 1.367 mm for input and output lines, respectively. In practice, though, most microwave connectors are designed for 50 Ω, so quarter-wavelength transformers are used to match the 100 Ω transmission lines to the 50 Ω output ports. The geometrical model of this T-junction power divider is presented in Figure 5-9.

The transmission coefficient magnitudes are shown in Figure 5-10 for this power divider as a function of frequency. Here an equal split is completely

FIGURE 5-9 ▪
Geometrical model
of a 3 dB T-junction
microstrip power
divider.

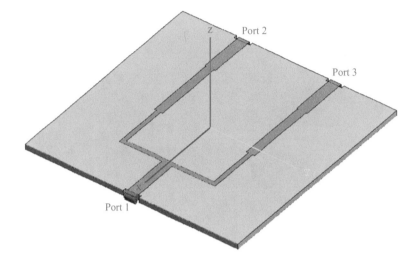

FIGURE 5-9 ▪
Geometrical model
of a 3 dB T-junction
microstrip power
divider.

FIGURE 5-10 ▪
Transmission
magnitude for the
3 dB T-junction
microstrip power
divider.

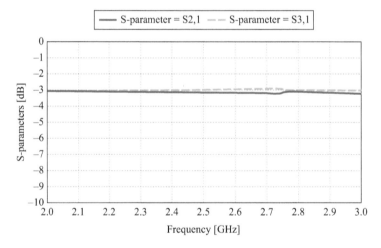

attained across the band. Figure 5-11 illustrates that the reflection coefficient at the input port is well matched across the band. The current distribution on the power divider is shown in Figure 5-12 at 2.45 GHz.

It is important to point out here that since the structure is completely symmetric the transmission coefficients $|S_{21}|$ and $|S_{31}|$ should be identical. The slight difference observed between these transmission magnitudes in Figure 5-10, however, is due to the asymmetry of the mesh used for the simulation. The subtle differences observed in the current distribution on the two arms of the junction in Figure 5-12 are also another indication of this. Improving the accuracy of the mesh usually alleviates these issues.

The reflection coefficients at all ports for this T-junction power divider are also outlined in Figure 5-13. Although the input port is well matched in

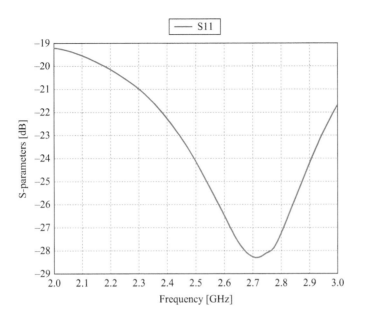

FIGURE 5-11 ■
Reflection coefficient at the input port for a 3 dB T-junction microstrip power divider.

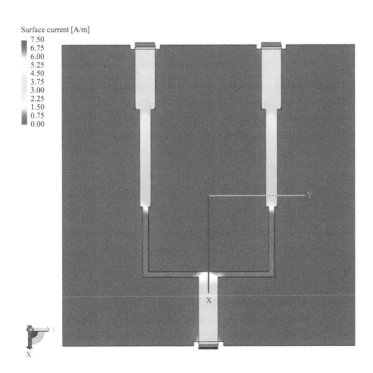

FIGURE 5-12 ■
Electric currents on a 3 dB T-junction microstrip power divider at 2.45 GHz.

FIGURE 5-13 ▪
Reflection coefficient
at the input and
output ports for a
3 dB T-junction
microstrip power
divider.

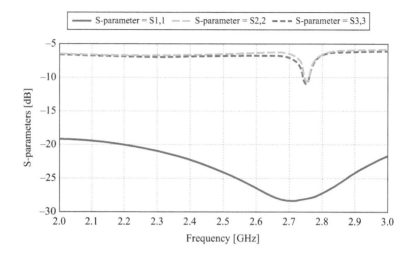

this design, the two output ports cannot achieve a good matching across the entire band. In general, this is the primary limitation of the T-junction power divider.

▌▌ 5.5 | WILKINSON POWER DIVIDERS

The T-junction power divider presented in the previous section suffers from two problems: it cannot achieve a good match at all ports (Figure 5-13), and it does not have good isolation between the output ports. The Wilkinson power divider is a three-port network that is theoretically perfectly matched at all ports and also provides good isolation between the output ports. As such, this design is very practical for microwave power dividers. Similar to the T-junction, it can achieve arbitrary power division, but again here we will consider only the equal split case. The design process here follows the approach given in [18].

For the Wilkinson power divider, all ports are attached to a transmission line with a characteristic impedance of Z_0. The input port is connected to the output ports with a transmission line with a characteristic impedance of $\sqrt{2}Z_0$ and a length of quarter-wavelength. The output ports are also joined together with a transmission line with a characteristic impedance of $2Z_0$, although usually a lumped component (e.g., a chip resistor) is used here.

The geometrical model of a Wilkinson power divider is shown in Figure 5-14. For this design, the transmission lines connecting the input and output ports are part of the circular ring. The ring microstrip line has a width of 2.716 mm

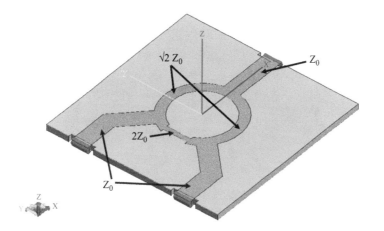

FIGURE 5-14 ▪
Geometrical model
of a microstrip
Wilkinson 3 dB
power divider.

(i.e., 70.71 Ω impedance). The transmission line connecting the two output ports has a characteristic impedance of 100 Ω.

Figure 5-15 demonstrates that the transmission magnitude can achieve an equal split of power (-3 dB). The electric surface current distribution on this microwave circuit at 2.45 GHz is shown in Figure 5-16.

5.6 | THE QUADRATURE HYBRID

The power dividers studied so far can realize different power ratios delivered at the two output ports; however, they cannot control the output phase between the

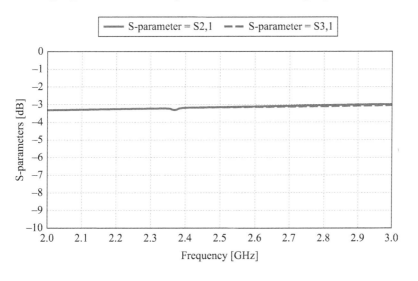

FIGURE 5-15 ▪
Transmission
magnitude of a
Wilkinson power
divider.

FIGURE 5-16 ▪

Electric surface
current distribution
on a Wilkinson
power divider at
2.45 GHz.

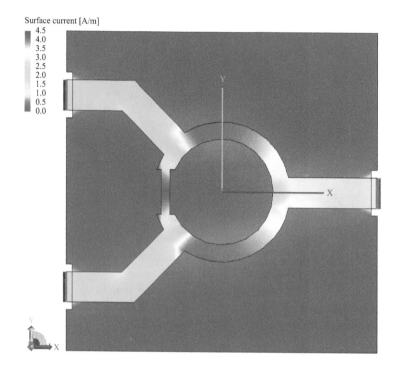

two ports. In many practical applications such as dual-port antennas, which are required to generate circular polarization, the two ports need an equal split of power and a phase difference of 90°. The latter can be accomplished using quadrature hybrids, which are usually called 3 dB directional couplers. Detailed design procedure and analysis of the scattering matrix for these microwave circuits are available in the literature [18].

Here we will show the performance of a quadrature hybrid designed using the same material at the design frequency of 2.45 GHz. The geometrical model of this microwave circuit is shown in Figure 5-17.

FIGURE 5-17 ▪

Geometrical model
of the quadrature
hybrid.

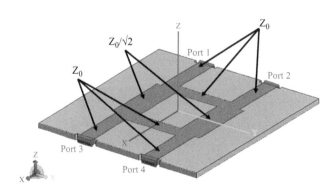

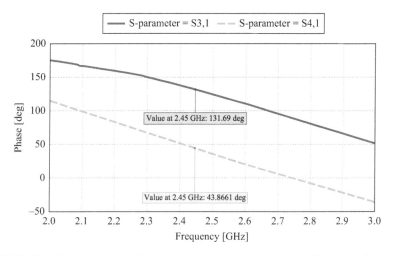

FIGURE 5-18 ▪ Transmission phase of a quadrature hybrid as a function of frequency.

The transmission phase at the output ports are shown in Figure 5-18. The phase difference between the two output ports is almost 88°, which is quite close to the ideal value of 90°.

The electric current distribution on the transmission lines is given in Figure 5-19, where the isolated port (port 2) is clearly showing a much lower surface current magnitude.

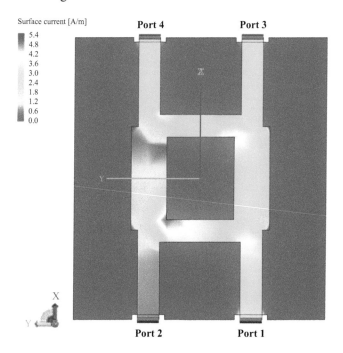

FIGURE 5-19 ▪ Electric currents on a quadrature hybrid at 2.45 GHz.

EXERCISES

(1) *Quarter-Wavelength Transformer.* Following the design procedure given in sections 5.2 and 5.3, design a quarter-wavelength microstrip transformer to match a 50 Ω input port to a 75 Ω output port at 5.8 GHz. Select a commercially available substrate.

(2) *Equal Split T-Junction Power Divider.* Using the same substrate selected in the first exercise and the design frequency of 5.8 GHz, design a power divider to realize equal power split at both output ports. For this design all ports are 50 Ω, so you would need to create a quarter-wavelength transformer for each output port as shown in section 5.3.

(3) *T-Junction Power Divider with 4:3 Ratio.* Using the same design properties as in exercise 2, design a T-junction power divider to achieve a power division of 4:3 between the output ports [18]. Compare the frequency behavior of these two designs. Which one achieves a better performance across the band?

Broadband Dipole Antennas

Chapter Outline

6.1 | INTRODUCTION

Chapter 2 covered wire dipole antennas and demonstrated that the current distribution on the wires is nearly sinusoidal. More importantly, the radiation characteristics of wire dipoles were shown to be very sensitive to frequency [7,8]. In other words, the thin wire dipole antennas studied earlier were extremely narrowband antennas. Many practical applications require a broad frequency band of coverage, though. In this chapter we will study some simple dipole type configurations that can provide a broader bandwidth: the cylindrical dipole and biconical antenna. In addition, the chapter introduces the folded dipole antenna, which can provide better matching characteristics than that of the single dipole. Design examples will also be presented for these antennas using the commercial software package FEKO [9].

6.2 | CYLINDRICAL DIPOLE ANTENNAS

6.2.1 Basics of the Cylindrical Dipole

The wire dipole antenna (Chapter 2) is a simple and inexpensive antenna with myriad applications. However, as discussed earlier, a very thin linear dipole has very narrowband input impedance characteristics. In general, the characteristics of a dipole antenna are frequency dependent, but the wire radius is proportional to the dipole bandwidth; that is, thick dipoles are considered broadband whereas thin dipoles are narrowband. A cylindrical dipole antenna is a simple wire dipole antenna with a thick wire diameter, and typically as the ratio of dipole length (l) over dipole diameter (d) (i.e., l/d) decreases the bandwidth increases.

From a computational analysis viewpoint, the primary difference is the dipole modeling and excitation. For thin wires, a simple wire model can be used for the dipole, and a wire port (with a voltage source) can provide the accurate excitation for this case. For cylindrical dipoles, though, the dipole has to be modeled as two separate cylinders. For the excitation, an edge port is defined between the two edges of the cylinders, which can give proper excitation for this configuration.

6.2.2 A Cylindrical Dipole Example

In this section, we will present the design and analysis of an ultra high frequency (UHF) cylindrical dipole antenna for a center operating frequency of 300 MHz. To model this dipole in FEKO, the upper and lower arms of the dipole are defined first as cylinders each having a diameter of 0.05λ. The two arms are separated with a distance of 0.01λ, and the total length of the dipole is 0.5λ. As discussed in Chapter 2, we would later tune the length of the dipole such that it resonates at the desired frequency. Another cylinder, which acts as the feed gap, is now defined between the two arms of the dipole with the same radius. All three cylinders are then united together. Now an edge port is defined between the upper arm of the dipole and the top of the feed cylinder. The geometrical model of the cylindrical dipole antenna in FEKO is given in Figure 6-1.

FIGURE 6-1 ▪
A cylindrical dipole antenna modeled in FEKO.

Usually the diameter of the feed cylinder does not have to equal the diameter of the dipole arms. In this study, though, these two diameters were made equal to better compare the results with the thin wire dipole models studied in Chapter 2.

As discussed earlier, in practice the antenna is fed with a 50 Ω transmission line; therefore, it is important to achieve a proper matching at the port. To this end, in the next stage the length of the dipole was tuned such that the antenna resonates at the design frequency of 300 MHz. The optimum length of the dipole was determined to be 0.435λ, and l/d for this design is 8.7, compared with the dipole example in section 2.5, where l/d was about 5,000. The simulated reflection coefficient for this cylindrical dipole antenna is given in Figure 6-2. In this illustration, the bandwidth obtained by this cylindrical dipole is significantly broader than the thin dipoles studied earlier. In particular, the -10 dB reflection bandwidth is about 45 MHz (15%) for the cylindrical dipole while this value was only 15 MHz (5%) for the thin dipole in Chapter 2.

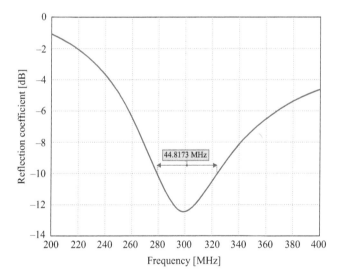

FIGURE 6-2 ▪
Reflection coefficient of a cylindrical dipole antenna.

The input impedance and gain of the cylindrical dipole as a function of frequency are presented in Figure 6-3 and Figure 6-4, respectively.

The results presented here clearly demonstrate the broadband characteristics of cylindrical dipole antennas. Importantly, the radiation characteristics of the cylindrical dipole are also similar to those of the wire dipole antenna. To illustrate this, the current distribution on the dipole at 300 MHz is shown in Figure 6-5. Similar to the wire dipoles, the current is maximum at the feed port

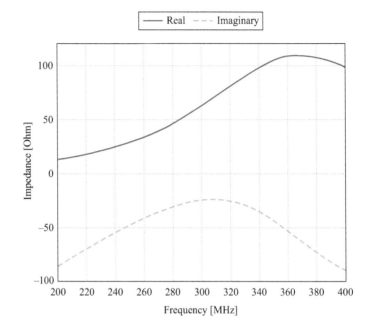

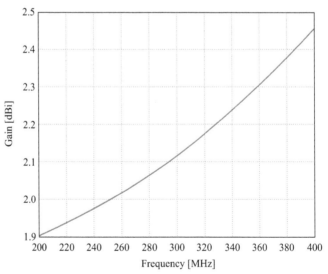

and decays toward the end of the dipole arms. The radiation pattern of the
cylindrical dipole is given in Figure 6-6 and Figure 6-7. As expected, the dipole
radiation pattern has a null along the z axis, and its maximum is in the
horizontal (x-y) plane.

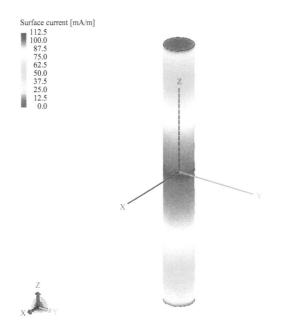

FIGURE 6-5 ▪ The electric currents on a cylindrical dipole antenna at 300 MHz.

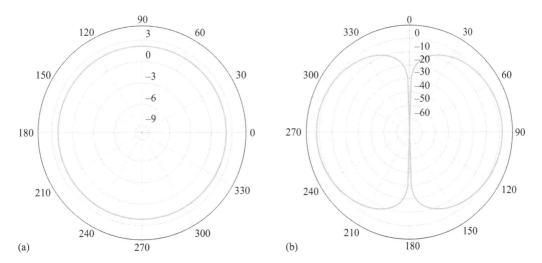

FIGURE 6-6 ▪ Polar plots of a cylindrical dipole antenna radiation patterns at 300 MHz: (a) x-y plane. (b) x-z plane.

FIGURE 6-7 ■ The
3D radiation pattern
of a cylindrical dipole
antenna at 300 MHz.

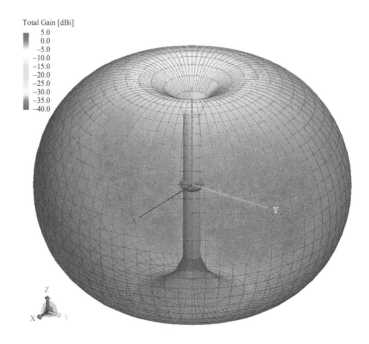

6.3 | BICONICAL ANTENNAS

6.3.1 Basics of the Biconical Antenna

The biconical antenna is a very practical and simple configuration that can also achieve broadband characteristics. It is a modified dipole shape that is basically similar to a cylindrical dipole except that the dipole's cylindrical arms are replaced with cones. A geometrical model of a finite size biconical antenna is pictured in Figure 6-8.

In the theoretical analysis it is generally assumed that the cones are infinite in extent [7,8], but in practice the cones have finite lengths. With the infinite model, the radiation characteristics can be computed assuming the dominant transverse electromagnetic (TEM) mode excitation as described in [7], and for infinite cones the input impedance is a direct function of the cone angle (α) and decreases as α is increased. The analysis of finite-length cones is more complicated, though, because some of the energy along the cone's surface is reflected and the remaining part is radiated. The finite-length biconical antenna becomes more broadband as the cone angle increases. Note that in Figure 6-8, the cone angle (α) is given with respect to the z-axis.

6.3.2 A Finite-Length Biconical Antenna Example

In this section we will study the performance of a finite-length biconical antenna designed for the operating frequency of 1 GHz. The biconical antenna

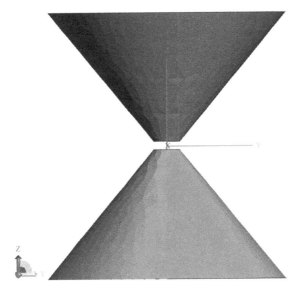

FIGURE 6-8 ▪ The basic geometry of a finite biconical antenna.

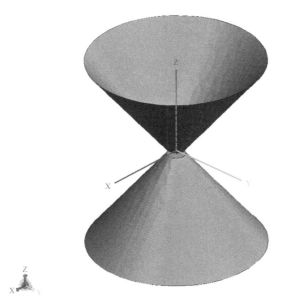

FIGURE 6-9 ▪ Geometry of a finite-length biconical antenna in FEKO.

parameters have been tuned to cover the frequency range from 0.5 to 2.0 GHz. The geometrical model of this biconical antenna in FEKO is given in Figure 6-9.

The biconical antenna has a length of 32 cm, and a feed gap of 1 cm is set between the two cones. The cones have an angle of 45.5°, and the radius of the lower base is 1.6 cm. Similar to the cylindrical dipole studied earlier, a cylinder is used to connect the two arms. The radius of this feed cylinder is 0.72 mm. From Figure 6-10, the biconical antenna is showing very broadband characteristics, much superior to the cylindrical dipole studied earlier. The simulated reflection coefficient for this biconical antenna is given in Figure 6-10. The −10 dB bandwidth for this design is from 0.65 GHz to 1.776 GHz (i.e., 2.7 approximately 3:1 bandwidth ratio).

FIGURE 6-10 ▪
Reflection coefficient
of a biconical
antenna.

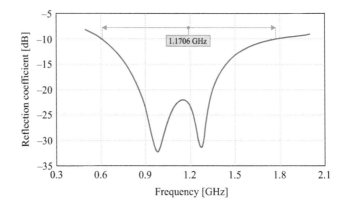

From Figure 6-10, the biconical antenna is showing very broadband characteristics, much superior to the cylindrical dipole studied earlier. For this design the bandwidth ratio is approximately 3:1. The input impedance and gain of this antenna as a function of frequency are given in Figure 6-11 and Figure 6-12, respectively. The broadband characteristics of this antenna are clearly demonstrated in the input impedance, where both real and imaginary parts show very desirable behavior across the band.

FIGURE 6-11 ▪
Input impedance of
a cylindrical dipole
antenna as a
function of
frequency.

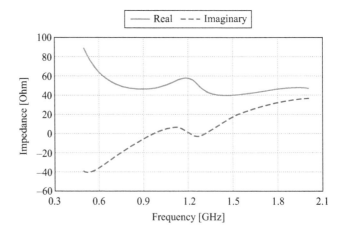

These results clearly demonstrate the very broadband characteristics of biconical antennas. As discussed earlier, biconical and dipole antennas have

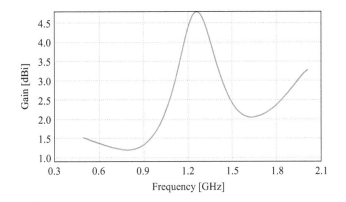

FIGURE 6-12 ▪
Gain of a cylindrical
dipole antenna as a
function of
frequency.

similar radiation characteristics. To illustrate this, the current distribution on the antenna at 1 GHz is presented in Figure 6-13, which shows that, similar to a dipole, the current is maximum at the feed port and decays toward the end of the dipole arms. This indicates that the antenna will also provide an omni-directional radiation pattern in the horizontal plane.

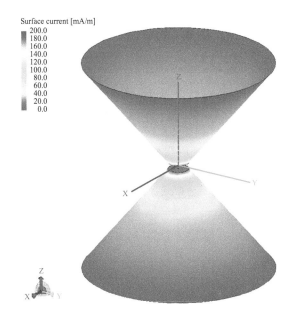

FIGURE 6-13 ▪
The electric currents
on a biconical
antenna at 1 GHz.

The radiation pattern of this biconical antenna is shown in Figure 6-14 and Figure 6-15, where as expected the pattern maximum is in the horizontal (x-y) plane.

FIGURE 6-14 ▪
Polar plots of a
biconical antenna
radiation patterns at
1 GHz: (a) x-y plane.
(b) x-z plane.

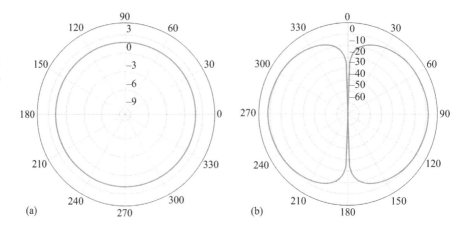

(a)

(b)

FIGURE 6-15 ▪ The
3D radiation pattern
of a biconical
antenna at 1 GHz.

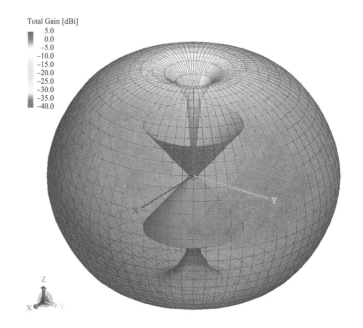

6.4 | FOLDED DIPOLE ANTENNAS

6.4.1 Folded Dipole Basics

To achieve good radiation characteristics, the antenna should provide good
matching to the transmission line feed of the antenna. The conventional $\lambda/2$
wire dipole antenna has an input impedance about $73 + j42.5 \ \Omega$. As such, the
dipole can easily be matched to conventional coaxial lines, which have characteristic impedances of 50 or 75 Ω. However, in practice several other types of

common transmission lines have characteristic impedance much higher than this. A good example is the twin-lead transmission line, which is widely used for TV applications and has a characteristic impedance of 300 Ω.

A folded dipole antenna is basically a very thin rectangular loop where the spacing between the two larger sides is very small (usually less than 0.05λ). A geometrical model of a folded dipole antenna is shown in Figure 6-16.

FIGURE 6-16 ■
Geometrical model
of a folded dipole.

A modified dipole configuration acting as a step-up impedance transformer, the folded dipole is a balanced system and as such is typically analyzed by its two modes [7]. In the transmission line mode, the currents on the two large sides of the loop are in opposite directions, whereas in the antenna mode these currents are in the same direction. Readers are encouraged to be familiar with the basic properties of these two modes [7,8].

The important characteristic of a half-wavelength folded dipole is that it has an input impedance equal to four times that of a dipole of the same length. Taking into account the impedance of a half-wavelength dipole, a $\lambda/2$ folded dipole is ideally suited for matching to a twin-lead transmission line. Moreover, the folded dipole has better bandwidth characteristics than a single dipole antenna of the same length.

6.4.2 Example of a Folded Wire Dipole

In this section we will study the performance of a folded wire dipole antenna designed for the operating frequency of 915 MHz. The antenna is to be matched to a transmission line with an input impedance of 300 Ω (i.e., a twin-lead transmission line). Similar to the dipole antennas studied earlier, the length of the dipole has to be tuned to provide matching at the design frequency. For this design, the spacing between the two larger sides is 0.025λ, and the wire radius is 0.001λ. The length of the dipole is determined to be 0.4525λ to provide optimum matching at 915 MHz.

The simulated reflection coefficient for this folded dipole antenna is given in Figure 6-17. A very good matching is obtained with this design at the center frequency of 915 MHz. Also, the −10 dB reflection bandwidth of this antenna is about 158 MHz (17%) which is significantly better than a conventional dipole antenna when matched to a 50 Ω transmission line. The input impedance of this folded dipole antenna is given in Figure 6-18.

FIGURE 6-17 ▪
Reflection coefficient of a folded dipole antenna.

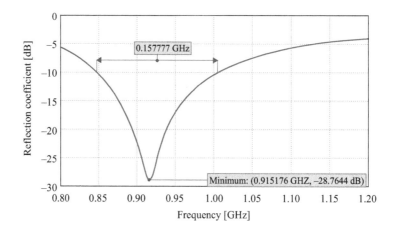

FIGURE 6-18 ▪
Input impedance of a folded dipole antenna.

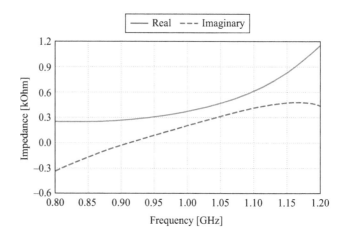

These results clearly demonstrate the broadband characteristic and the impedance transformer property of folded dipoles. As discussed earlier, the impedance of this folded dipole is four times greater than the half-wavelength dipole, or about 300 Ω. The gain of this folded dipole as a function of frequency is also given in Figure 6-19. Again, the spike observed here is due to the numerical errors due to using the CADFEKO interpolation option.

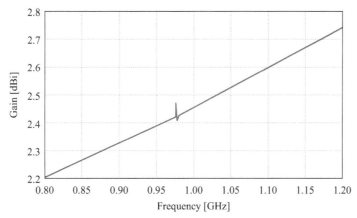

FIGURE 6-19 ▪
Gain of a folded dipole antenna as a function of frequency.

The gain of the folded dipole antenna in the figure is about the same as the single dipole. This antenna also has similar radiation characteristics as a dipole antenna, which can best be illustrated by observing the current distributions on the antenna. The electric currents on the antenna at the operating frequency of 915 MHz is shown in Figure 6-20, which illustrates a dipole-type current on the

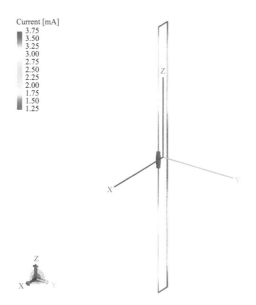

FIGURE 6-20 ▪
Currents on a folded dipole antenna at center design frequency.

larger sides of the loop. The currents on the smaller sides of the loop are almost zero, which ultimately would result in a dipole type of radiation pattern for the folded dipole antenna. A model showing the location of the vertices of the mesh elements is given in Figure 6-21. Figure 6-22 presents a 2D plot of the currents

FIGURE 6-21 ■
The geometry of a folded dipoles and the location of the mesh vertices on the wires.

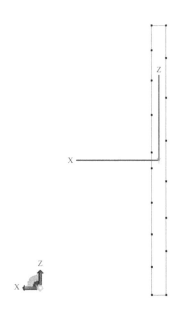

FIGURE 6-22 ■
Currents on the wire segments.

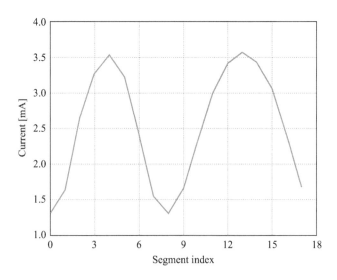

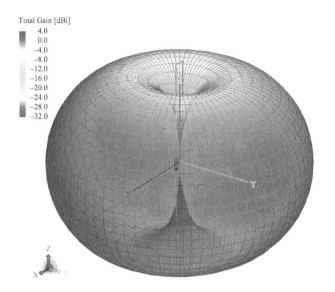

Total Gain [dBi]
4.0
0.0
−4.0
−8.0
−12.0
−16.0
−20.0
−24.0
−28.0
−32.0

FIGURE 6-23 ▪
Radiation pattern of
a folded dipole
antenna at 915 MHz.

on the segments of the wires and shows that the peak current is almost identical on both wires.

As discussed earlier, with such a current distribution it is expected that the folded dipole will have a dipole-type radiation pattern. This is indeed the case, and the 3D pattern of this folded dipole antenna at 915 MHz is shown in Figure 6-23.

EXERCISES

(1) *Cylindrical Dipole Antennas.* Following the design example given in section 6.2, study the effect of increasing the radius of the cylinder on the bandwidth of the antenna. Note that in each case as the radius increases one would also need to tune the length of the dipole to ensure that it resonates at the center frequency. Does increasing the radius improve the bandwidth? Explain your answer.

(2) *Hemispherical Dipole Antennas.* A hemispherical dipole antenna is one where the cylindrical arms of the dipole are replaced by hemispherical conductors. Study the performance of this dipole for the case when the radius of the hemisphere is 0.25λ. A CADFEKO model of this antenna is shown in Figure P6-1. Does this design achieve a better bandwidth than the cylindrical dipole?

FIGURE P6-1 ▪
A hemispherical
dipole antenna.

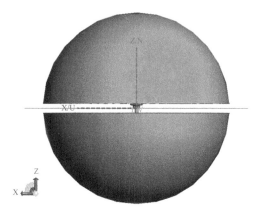

(3) ***Conical Antennas.*** Using the same design dimension of the biconical antenna presented in section 6.3, design a conical (unipole) antenna. In a conical antenna, one of the cones is replaced by an infinite ground plane, which basically makes it a monopole. Compare the performance of this antenna with the biconical design presented in this chapter.

(4) ***Biconical Antenna with a Spherical Cap.*** One approach to improve the bandwidth of the biconical antenna is to add a spherical cap on both ends of the bicone. The geometrical model of a biconical antenna with a spherical cap is shown in Figure P6-2. The radius of the generating sphere is equal to the distance from the cone apex to the flat base of the cone. Study the performance of the biconical antenna presented in section 6.3 by adding this spherical cap. Does this antenna achieve a wider bandwidth? Why?

FIGURE P6-2 ▪
A biconical antenna
with a spherical cap.

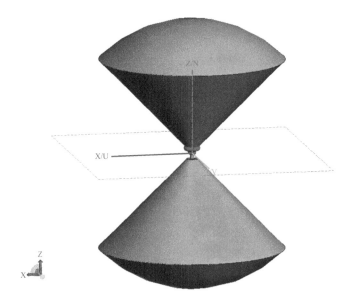

(5) ***Bowtie Antennas.*** Printed antennas are desirable in many applications due to their low profile. A printed dipole antenna can achieve a similar performance as the wire dipoles studied in Chapter 2, with the added advantage of simpler fabrication. A half-wavelength printed dipole antenna is shown in Figure P6-3.

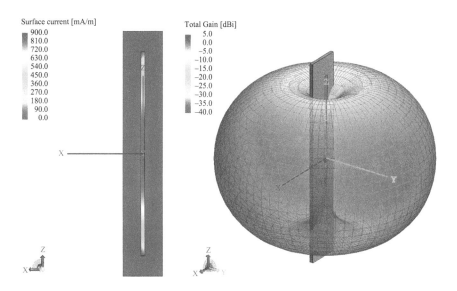

FIGURE P6-3 ▦
A printed half-wavelength dipole antenna and its 3D radiation pattern.

Similar to the wire dipoles, the major problem is that the printed dipole also has a narrow bandwidth. On the other hand, a bowtie antenna is a printed dipole configuration, where the rectangular arms of the dipole are replaced by triangular sheets. A bowtie antenna is shown in Figure P6-4.

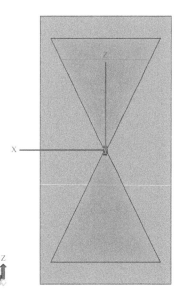

FIGURE P6-4 ▦
A printed bowtie antenna.

Similar to the biconical antenna, the most important parameter for a bowtie is the half-angle of the triangular sheet. Design a bowtie antenna for the operating frequency of 2.45 GHz, and study the effect of the geometrical parameters on the performance of the antenna. Can this antenna achieve a wider bandwidth than the printed dipole? Explain your answer.

(6) **Folded Dipole Antennas.** Design a UHF half-wavelength folded dipole antenna for the operating frequency of 300 MHz. Compare the performance of this design with the UHF dipole antenna designed in Chapter 2, and verify that the input impedance of the folded dipole is four times larger than that of the dipole antenna.

Traveling Wave and Broadband Antennas

7.1 | INTRODUCTION

Conventional center-fed linear wire or cylindrical type antennas were studied in the previous chapters. These types of configurations have an almost sinusoidal current amplitude distribution, with an approximately constant phase distribution. However, as the length of the wire increases beyond the conventional $\lambda/2$ dipole, the sinusoidal current distribution becomes a standing wave as a result of the reflection at the open end of the wire. This chapter will study two very practical traveling wave antennas: helical and Yagi-Uda. The helical antenna is a very good example of a traveling wave antenna that exhibits excellent radiation characteristics. It consists of a single or multiple conductors that are wound into a helical shape. It has broadband characteristics and also can provide circular polarization, which is a desired feature for many applications. The Yagi-Uda antenna consists of a linear array of elements with only one of the elements energized. The other elements in the array act as parasitic elements, and their currents are induced by mutual coupling. This chapter will review the basic theories of these two antennas, and several design examples will be presented.

▌ 7.2 | HELICAL ANTENNAS

7.2.1 Basic Theory and Operating Modes

In this section we will study one of the most popular circularly polarized broadband antennas. A basic, simple, and practical configuration of an electromagnetic radiator is that of a conducting wire wound in the form of a screw thread, forming a helix. In most cases, the helix is mounted on a ground plane. The ground plane can take different forms, but the conventional type is a flat circular plane, where typically the diameter of the ground plane should be at least $3\lambda/4$.

The geometrical configuration of a helix usually consists of N turns, diameter (D), and spacing (S) between any two adjacent turns. A geometrical model of a helix antenna, along with these design parameters, is given in Figure 7-1a. The starting point of the helix wire is not the geometric center of the circular ground; however, the helix is placed at the center of the ground plane, as shown in Figure 7-1b.

FIGURE 7-1 ▪
Geometry of a helix
antenna: (a) 3D view.
(b) Top view.

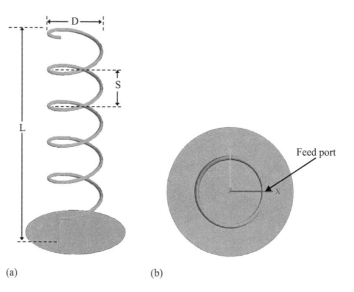

(a)   (b)

One other important geometrical parameter for the helix is the pitch angle, which is formed by a line tangent to the helix wire and plane perpendicular to the helix axis. For better illustration, a geometrical model of one uncoiled turn of a helix is shown in Figure 7-2, where L_0 is the length of single turn. Mathematically this is given by

$$\alpha = \tan^{-1}\left(\frac{S}{\pi D}\right). \tag{7-1}$$

The antenna's radiation characteristics can be varied by controlling the size of its geometrical properties compared with wavelength. Its general polarization

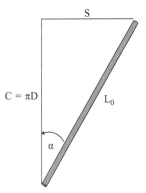

FIGURE 7-2 ▪ The pitch angle of a helix.

is elliptical, but circular and linear polarizations can be achieved over different frequency ranges. The helical antenna can operate in many modes; however, the two principal ones are the normal (broadside) and the axial (end-fire) modes. The normal mode has its maximum in a plane normal to the axis of the helix and is nearly null along the axis. The pattern is similar in shape to that of a small dipole or circular loop. The axial mode has its maximum along the axis of the helix and is similar to that of an end-fire array. The axial (end-fire) mode is usually the most practical because it can achieve circular polarization over a wider bandwidth (usually 2:1) with good efficiency.

To achieve the normal mode of operation, the dimensions of the helix are usually small compared with the wavelength. In the normal mode, circular polarization ($AR = 1$) can be achieved by having [7]

$$\tan \alpha = \frac{S}{\pi D} = \frac{\pi D}{2\lambda_o}. \tag{7-2}$$

Under this condition, the helix will have a circularly polarized radiation in all directions other than broadside. As discussed earlier, though, because the dimensions of the normal mode helix have to be very small compared with wavelength this mode also has a poor efficiency and a very narrow bandwidth and is therefore seldom used.

The practical mode of operation for the helix is the axial mode. In this mode, only one major lobe is generated, and it is in the end-fire direction. To achieve this, the diameter and spacing have to be large relative to the wavelength. Empirical formulas have been obtained for this mode of operation and are used to determine these parameters [7,8,19]. These design guidelines are summarized as follows:

$$3/4 \leq C/\lambda_o \leq 4/3 \quad \text{(with } C/\lambda_o = 1 \text{ considered optimum value)}, \tag{7-3}$$

$$S \approx \lambda_o/4, \tag{7-4}$$

$$12° \leq \alpha \leq 14°. \tag{7-5}$$

7.2.2 Full-Wave Simulation of Helical Antennas in FEKO

To simulate a helix in FEKO, we set up the simulation problem in the CADFEKO design environment. Designing a helix in FEKO is relatively simple since a helix option is directly available in the "Create arc" tab. A snapshot of the create helix screen is given in Figure 7-3.

FIGURE 7-3 ■
Designing a helix
in CADFEKO.

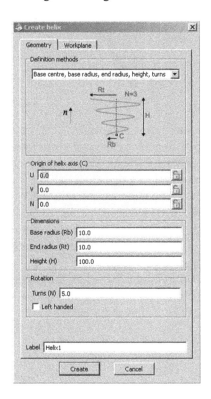

The conventional helix is designed with equal radius for all the turns, but there is an option to taper the helix's winding. The other necessary design components are a feed wire line and a circular ground plane. Once these three components are designed, they must be united to ensure a correct meshing of the antenna before starting the simulation. In the next section we will study an ultra high frequency (UHF) helical antenna designed for normal and axial mode operations.

7.2.3 The Normal Mode Helix

In the normal mode of operation, the field radiated by the antenna is maximum in a plane normal to the helix axis and minimum along its axis. To achieve the normal mode of operation, the dimensions of the helix should be small compared with wavelength. Here we select a two-turn helix with a total length of 0.2λ. The diameter of the ground plane is set to 0.25λ. A wire with a length of 0.001λ is used to connect the helix to the ground plane. The geometry of this helix is presented in Figure 7-4.

The gain pattern of the helix is given in Figure 7-5. For this operating mode, the maximum radiation is normal to the direction of the helix axis. It should be noted that this normal mode helix is designed for linear polarization.

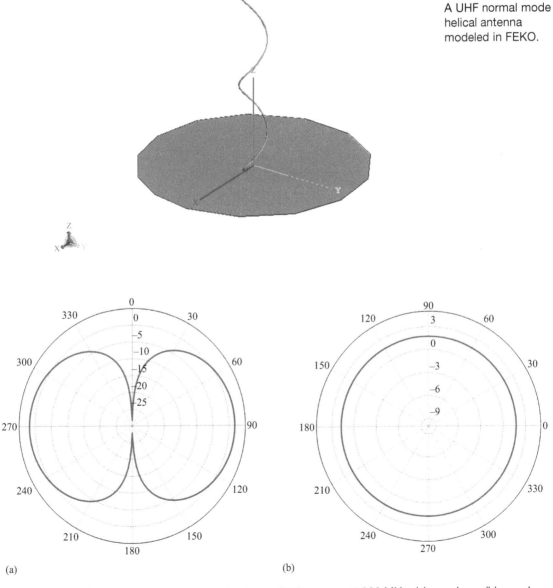

FIGURE 7-4 ■
A UHF normal mode helical antenna modeled in FEKO.

(a)

(b)

FIGURE 7-5 ■ The gain pattern of a normal mode helical antenna at 900 MHz: (a) x-z plane. (b) y-z plane. (c) 3D pattern.

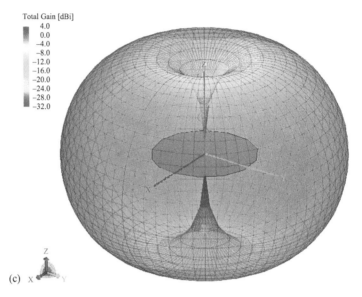

FIGURE 7-5 ▪ *(continued)*

The antenna gain and the input impedance as a function of frequency are illustrated in Figure 7-6. In addition to a very low gain performance, the normal mode helix does appear to have a very narrow operating band. In fact, because of the critical dependence of its radiation characteristics on its geometrical dimensions, which must be very small compared with the wavelength, this mode of operation is very narrow in bandwidth and its radiation efficiency is very low. In practice, due to these limitations this operating mode of the helix is seldom utilized.

7.2.4 The Axial Mode Helix

As discussed earlier, the practical operating mode of the helix antenna is the axial or end-fire mode. To study the performance of an axial mode helix, here we select a seven-turn helix with a total length of 1.61λ designed for the operating frequency of 300 MHz. The radius of the helix is set to $\lambda/2\pi$ [7], and the diameter of the ground plane is set to 0.6λ. A wire with a length of 0.15λ is used to connect the helix to the ground plane. The geometry of this helix is given in Figure 7-7. The helix is positioned at the center of the circular ground plane such that from the top view the center of the ground and helix are concentric.

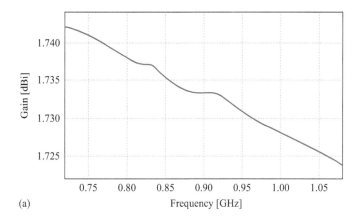

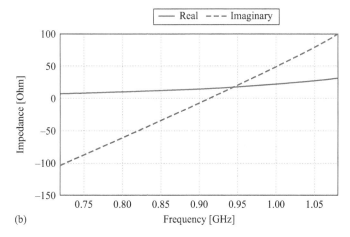

(a)

(b)

FIGURE 7-6 ▪
(a) Gain versus
frequency for a helix.
(b) Real and
imaginary
components of input
impedance versus
frequency.

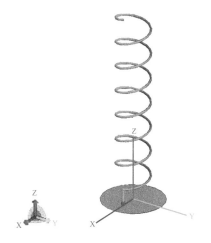

FIGURE 7-7 ▪
A UHF axial mode
helical antenna
modeled in FEKO.

The current distribution on the wire and the directivity pattern of the helix are shown in Figure 7-8. For this operating mode, the beam direction is along the axis of the helix.

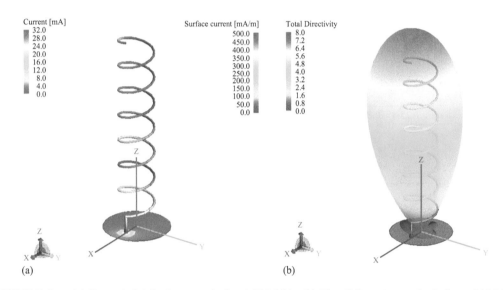

FIGURE 7-8 ▪ (a) Current distribution on a helix at 300 MHz. (b) Directivity pattern of a helix at 300 MHz.

The helical antenna is designed to achieve a right-hand circularly polarized radiation performance. The antenna gain as a function of frequency is given in Figure 7-9a. The cross-polarized gain (left-hand circularly polarized) is at least 15 dB lower in the operating band. The real and imaginary parts of the impedance are also presented in Figure 7-9b. The impedance is almost stable across the shown frequency range, illustrating the wide band nature of the helical antenna. In general, the input impedance is critically dependent on the pitch angle and the size of the conducting wire, especially near the feed point, and it can be adjusted by controlling their values. In practice, however, where the antenna is fed with a 50 Ω transmission line some further modifications are required [7,8].

7.2.5 Various Ground Shapes for Helical Antennas

Different ground shapes have been proposed for the helix, one of the most practical of which is the cupped ground plane in the form of a cylindrical cavity or a frustrum cavity. Here we will study the cupped ground plane based on the dimensions given by Kraus [19]. The geometrical model of this antenna in FEKO is given in Figure 7-10. The height of the cylindrical cavity is 3/8 of the wavelength at center frequency.

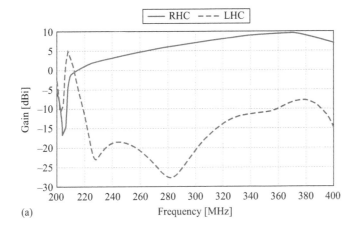

(a)

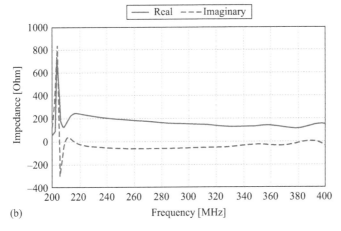

(b)

FIGURE 7-9 ▪
(a) Gain versus frequency for a helix. (b) Real and imaginary components of input impedance as a function of frequency.

FIGURE 7-10 ▪
A UHF helical antenna modeled in FEKO (Kraus design).

The radiation pattern of this helix is shown in Figure 7-11. A much higher directivity (i.e. 12.5, compared with the previous design where the directivity was 8.0) is obtained. Moreover, the slight tilt in the pattern is corrected, and the beam is exactly in the end-fire direction.

FIGURE 7-11 ■
Radiation pattern of a helix at 300 MHz: (a) Total directivity. (b) Right-hand circularly polarized gain in dB.

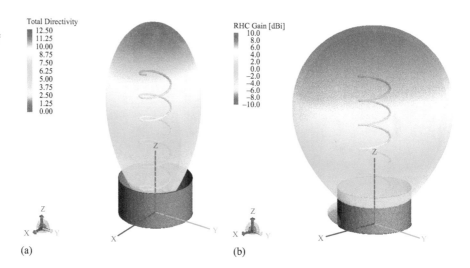

(a)   (b)

The co- and cross-polarized radiation patterns of this helix at 300 MHz are given in Figure 7-12a. This design achieves a very low-cross polarization. The axial ratio in the end-fire direction as a function of frequency is illustrated in Figure 7-12b. The helix maintains its low-cross polarized feature over the operating band.

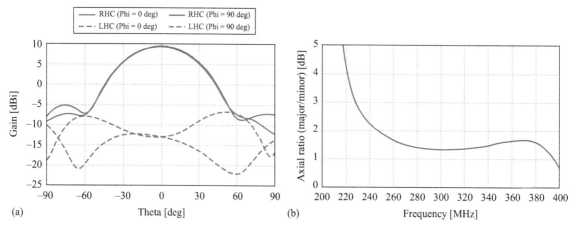

FIGURE 7-12 ■ (a) Radiation pattern of a helix antenna at 300 MHz. (b) Axial ratio of a helix versus frequency.

The antenna gain as a function of frequency is given in Figure 7-13a. Similar to the previous design, the cross-polarized gain (left-hand circularly polarized) is at least 15 dB lower in the operating band. The real and imaginary parts of the impedance are presented in Figure 7-13b. As discussed already, one practical challenge in designing the helix is matching it to the 50 Ω transmission line.

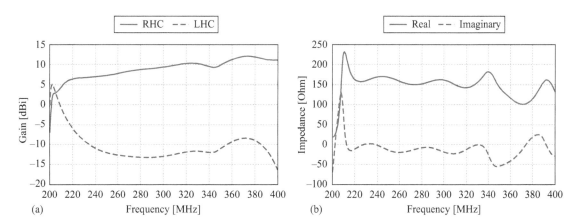

FIGURE 7-13 ▪ (a) Gain versus frequency for a helix. (b) Real and imaginary components of input impedance as a function of frequency.

7.3 | YAGI-UDA ANTENNAS

7.3.1 Basics

It is well known from antenna theory that array antennas can be used to increase directivity. However, an array with all active elements does require a direct connection to each element by a feed network. On the other hand, if the number of fed elements is minimized, the feed network can be simplified to a great extent. In general, these types of arrays, where all elements are not active, are known as parasitic arrays [7,8]. A parasitic linear array of parallel dipoles is typically known as a Yagi-Uda antenna or array. The basic operating principle of a Yagi-Uda antenna relies on using parasitic directors and reflectors to direct the main beam of the antenna in the desired direction. A very illustrative theoretical discussion using array theory is provided in [8], and readers are encouraged to study this to gain better insight into the fundamental theory of parasitic elements. Here, though, we will demonstrate the effects of reflector and director elements on the performance of a dipole antenna.

Consider a two-element linear dipole array where the dipoles are spaced at a distance of 0.04λ apart in the xz plane. The driven element has a length of 0.4781λ, and the parasitic element has length of 0.49λ. Both elements have a wire radius of 0.001λ. The geometrical model of this two-element dipole array along with the current distribution on the wires is shown in Figure 7-14. The driven element is at the center of coordinate system in the x-y plane.

FIGURE 7-14 ▪

A two-element dipole array consisting of a driven element and a reflector.

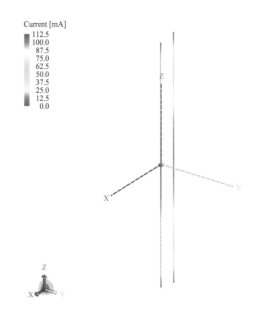

Due to the coupling effects between the two elements, a strong electric current is also present on the parasitic element. The radiation pattern of this dipole array is shown in Figure 7-15 and Figure 7-16.

The results shown here demonstrate that the slightly larger parasitic element is directing the main beam of the array in the end-fire direction. Since the parasitic element is directing the beam toward the positive x axis, it is known as a reflector. Also note that the length of the reflector is larger than the driven element.

Now let us consider a two-element linear dipole array where again the dipoles are spaced at a distance of 0.04λ apart and the driven element has a length of 0.4781λ. Also, both elements have a wire radius of 0.001λ. The parasitic element, however, now has a length of 0.45λ. The geometrical model of this two-element dipole array along with the current distribution on the wires is presented in Figure 7-17. Again, the driven element is at the center of coordinate system in the x-y plane, but the parasitic element is now placed in front of the driven element in the x direction.

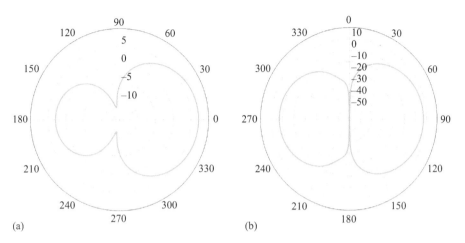

FIGURE 7-15 ■ Polar plots of a two-element array radiation patterns: (a) x-y plane. (b) x-z plane.

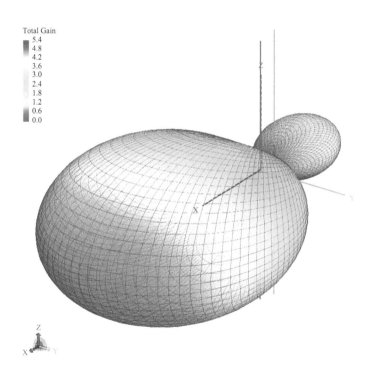

FIGURE 7-16 ■ 3D radiation pattern of the two-element dipole array.

Similarly, due to the coupling effects between the two elements, a strong electric current is also present on the parasitic element. The radiation pattern of this dipole array is shown in Figure 7-18 and Figure 7-19. These results show that this slightly smaller parasitic element is also directing the main beam of the

FIGURE 7-17 ▪
A two-element
dipole array
consisting of a
driven element
and a director.

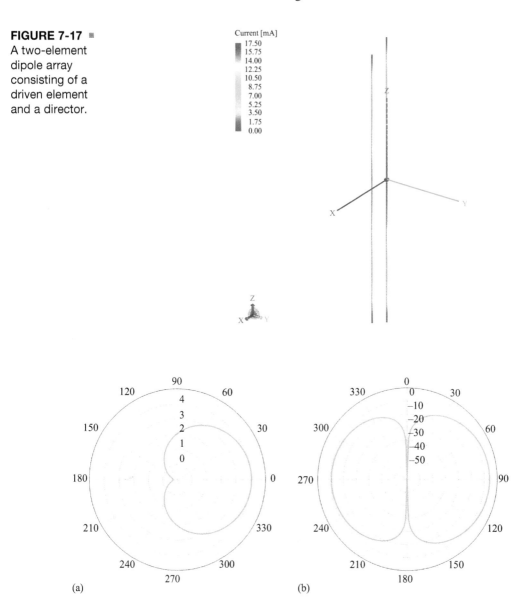

FIGURE 7-18 ▪ Polar plots of a two-element array radiation patterns: (a) x-y plane. (b) x-z plane.

array in the end-fire direction, so it is called the director. The length of the
director is smaller than the driven element.

The end-fire beams generated by the reflector and director alone are the key in
realizing a high-gain end-fire radiation from a Yagi-Uda antenna array. Adding
more parasitic elements can increase the array gain, but in most cases only one
reflector element is sufficient and the number of director elements is increased to

Total Gain
2.7
2.4
2.1
1.8
1.5
1.2
0.9
0.6
0.3
0.0

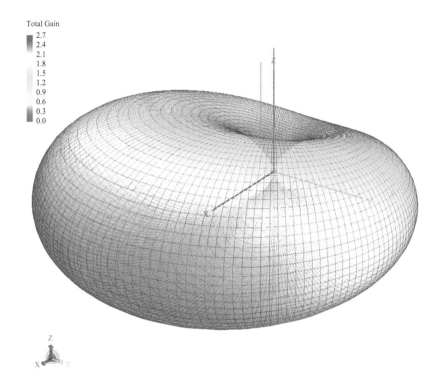

FIGURE 7-19 ■ 3D radiation pattern of a two-element dipole array.

achieve a higher gain. Due to the strong effects of mutual coupling between the elements of a Yagi-Uda antenna array, direct design equations are not available, but some design guidelines have been derived [7,8]. In the next sections, we will demonstrate the performance of several different design examples.

7.3.2 A Three-Element Yagi-Uda Array

A simple three-element array, consisting of one driven element and two parasitic elements, is demonstrated here. The dipoles are spaced at a distance of 0.04λ apart and all have a wire radius of 0.001λ. The driven element has a length of 0.4781λ. The reflector and director elements have a length of 0.49λ and 0.45λ, respectively. The geometrical model of this Yagi-Uda dipole array along with the current distribution on the wires is shown in Figure 7-20.

The radiation pattern of this dipole array is illustrated in Figure 7-21 and Figure 7-22.

The simple three-element configuration clearly reveals the effective use of parasitic elements in Yagi-Uda arrays. The gain of this antenna is 7.2 dBi, which is a significant improvement compared with an isolated dipole antenna. As discussed earlier, further gain enhancement is possible by increasing the number of directors in the array.

FIGURE 7-20 ■
A three-element
Yagi-Uda dipole
array.

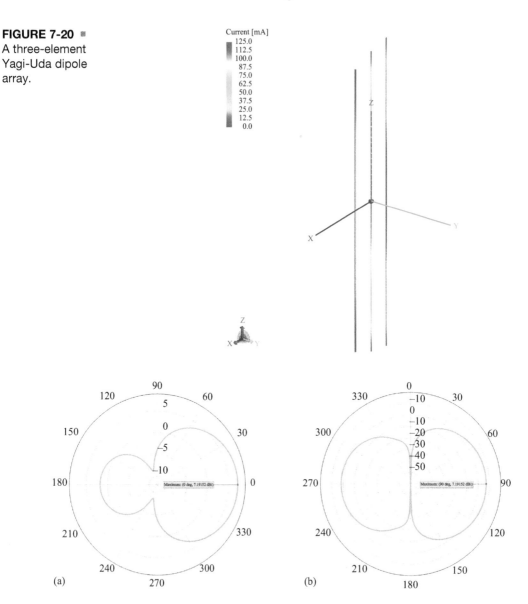

FIGURE 7-21 ■ Polar plots of a three-element Yagi-Uda dipole array radiation patterns: (a) x-y plane. (b) x-z plane.

These results demonstrate that the slightly larger parasitic element is directing the main beam of the array in the end-fire direction. Since the parasitic element is directing the beam toward the positive x axis, this element is known as a reflector. The length of the reflector is also larger than the driven element.

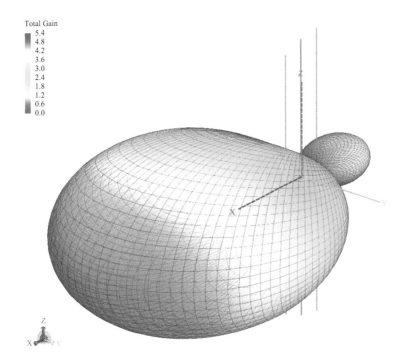

Total Gain
5.4
4.8
4.2
3.6
3.0
2.4
1.8
1.2
0.6
0.0

FIGURE 7-22 ▪ 3D radiation pattern of a three-element Yagi-Uda dipole array.

7.3.3 A Fifteen-Element Yagi-Uda Array

To demonstrate that the gain of Yagi-Uda antennas can be increased by increasing the number of directors, here we study a fifteen-element array, consisting of one driven element, one reflector element, and thirteen director elements [7]. The driven element has a length of 0.47λ, and the reflector and all director elements have a length of 0.5λ and 0.406λ, respectively. The spacing between the reflector and feeder is 0.25λ, and spacing between all adjacent directors is 0.34λ. All element of this Yagi-Uda dipole array have a wire radius of 0.003λ. The geometrical model of this Yagi-Uda dipole array along with the current distribution on the wires is shown in Figure 7-23.

The radiation pattern of this dipole array is shown in Figure 7-24 and Figure 7-25.

The gain of this fifteen-element antenna array is about 14.3 dBi. As discussed earlier, this can be further increased by increasing the number of directors; however, a saturation effect will be observed beyond a certain limit depending on the design, and the majority of Yagi-Uda antennas have about six to twelve director elements.

One other issue is regarding the bandwidth of Yagi-Uda arrays. In general, these antennas are resonant type structures and cannot achieve broad bandwidths.

FIGURE 7-23 ◾
A fifteen-element
Yagi-Uda dipole
array.

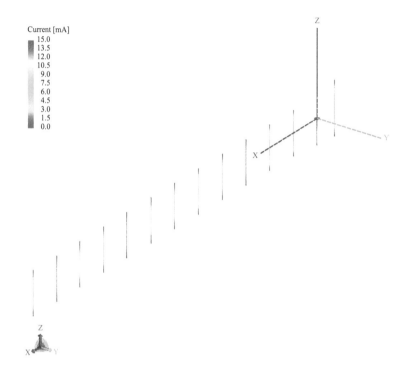

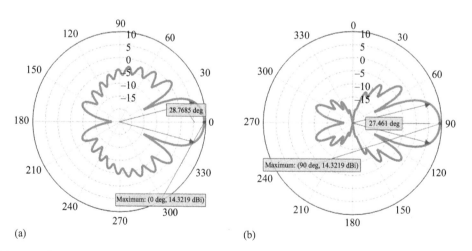

(a)   (b)

FIGURE 7-24 ◾ Polar plots of a fifteen-element Yagi-Uda dipole array radiation patterns: (a) x-y plane. (b) x-z plane.

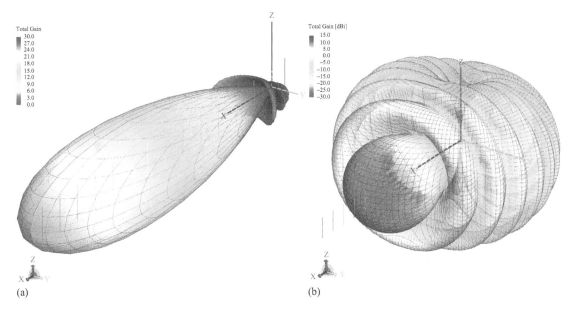

(a) (b)

FIGURE 7-25 ▪ 3D radiation pattern of a fifteen-element Yagi-Uda dipole array: (a) Total gain. (b) Total gain in dBi.

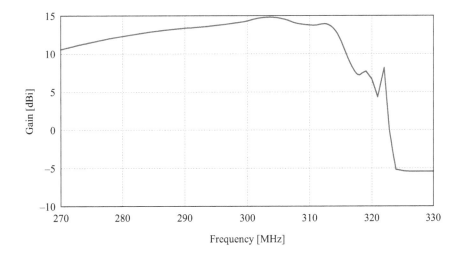

FIGURE 7-26 ▪ Gain as a function of frequency for a fifteen-element Yagi-Uda dipole array.

The gain and reflection coefficient as a function of frequency are given in Figure 7-26 and Figure 7-27 for this array. The center frequency for this design was 300 MHz, but all dimensions are given in wavelengths so this design can be rescaled if necessary.

FIGURE 7-27 ▪
Reflection coefficient (with a 50 Ω port impedance) as a function of frequency for a fifteen-element Yagi-Uda dipole array.

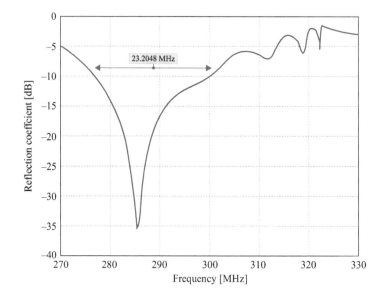

7.3.4 An Optimized Six-Element Yagi-Uda Array

In the design examples presented earlier, the length of all director elements and their adjacent spacing were held constant. Numerous studies have shown that the radiation characteristics of the array can be adjusted by controlling the geometrical parameters of the array. Given the fact that no design equation really exists for a Yagi-Uda array, a conventional choice for the design is to use optimization techniques. Here we will show the performance of an optimized Yagi-Uda dipole array, where all element lengths and director spacings have been optimized. The optimized dimensions of this six-element Yagi-Uda array are listed in Figure 7-28.

Directivity optimization for 6-element Yagi-Uda array (a = 0.003369λ)

	l_1/λ	l_2/λ	l_3/λ	l_4/λ	l_5/λ	l_6/λ	S_{21}/λ	S_{32}/λ	S_{43}/λ	$S_{54}\lambda$	S_{65}/λ	**Directivity**
Initial Array	0.510	0.490	0.430	0.430	0.430	0.430	0.250	0.310	0.310	0.310	0.310	10.93 dB
After Spacing Perturbation	0.510	0.490	0.430	0.430	0.430	0.430	0.250	0.298	0.406	0.323	0.422	12.83 dB
After Spacing & Length Perturbation	0.472	0.452	0.436	0.430	0.434	0.430	0.250	0.298	0.406	0.323	0.422	13.41 dB

FIGURE 7-28 ▪ Optimized parameters of a six-element Yagi-Uda dipole array [7].

The geometrical model of this optimized Yagi-Uda dipole array along with the current distribution on the wires is shown in Figure 7-29, where *l* is the length of each dipole, and *S* is the dipole separation, as shown in Figure 7-29.

The radiation pattern of this dipole array is shown in Figure 7-30 and Figure 7-31.

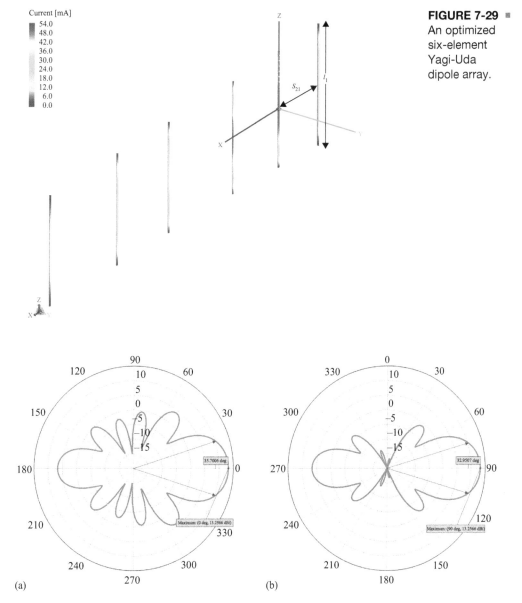

FIGURE 7-29 ▪ An optimized six-element Yagi-Uda dipole array.

FIGURE 7-30 ▪ Polar plots of an optimized six-element Yagi-Uda dipole array radiation patterns: (a) x-y plane. (b) x-z plane.

FIGURE 7-31 ▪
3D radiation pattern
of an optimized
six-element
Yagi-Uda dipole
array.

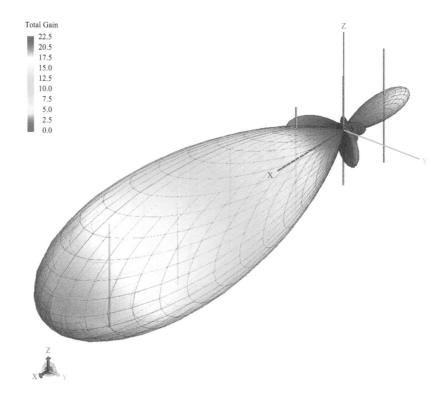

The gain of this optimized six-element antenna array is about 13.25 dBi, which is quite close to the value reported in [7]. The gain and input impedances as a function of frequency are given in Figure 7-32 and Figure 7-33 for this array. Similar to the fifteen-element array, the center frequency for this design was set to 300 MHz.

FIGURE 7-32 ▪
Gain as a function
of frequency for
an optimized
six-element
Yagi-Uda dipole
array.

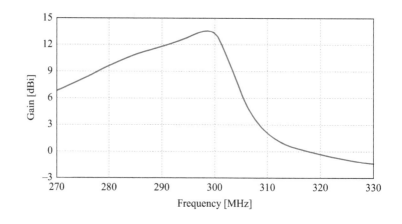

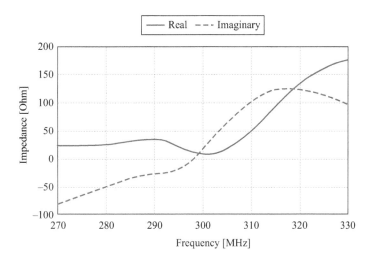

FIGURE 7-33 ▪
Input impedance as a function of frequency for an optimized six-element Yagi-Uda dipole array.

EXERCISES

(1) ***Designing an Axial Mode Helix.*** Design an axial mode helix antenna for the operating frequency of 5 GHz. Study the effect of the number of turns (N) on the gain and axial ratio of the antenna. Does the performance of the helix improve as the number of turns is increased? Justify your answer.

(2) ***Parametric Studies for Kraus's Helix Model.*** For the configuration given in Figure 7-9, change the size of the ground plane and height of the cavity walls and study the effect of these parameters on the input impedance and radiation pattern of the antenna.

(3) ***A Cone-Shaped Ground Model for Helix.*** Change the cylindrical ground studied in exercise 2 to a cone-shaped ground. The geometry of this configuration is shown in Figure P7-1. Study the effect of cone parameters on the performance of the helix. Can this ground achieve a better performance than the cylindrical type? Explain your answer.

(4) ***Dielectric Loading of Helix.*** For the configuration given in Figure 7-9, insert a dielectric cylinder inside the helix and study the gain, directivity, and input impedance for dielectric constants of 1 to 10.

(5) ***Designing a Helical Antenna Array.*** Using the axial mode UHF helical antenna studied in section 7.2 as the element, design a 2×2 array antenna as shown in Figure P7-2. Assume that the impedance of each of the element ports is 200 Ω. Excite all ports with a voltage source magnitude of 1 and study the effect of port phase on the performance of the helical array. Assign a phase excitation for the feed ports such that it improves the polarization purity (for circular polarization).

FIGURE P7-1 ■
A cone-shaped
ground plane for
helix antenna.

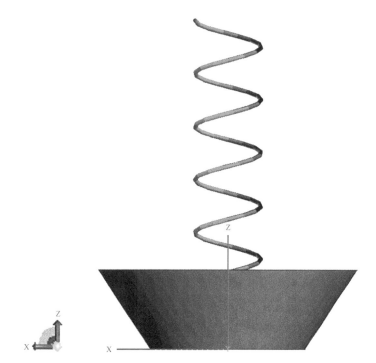

FIGURE P7-2 ■
A four-element
helical array
antenna.

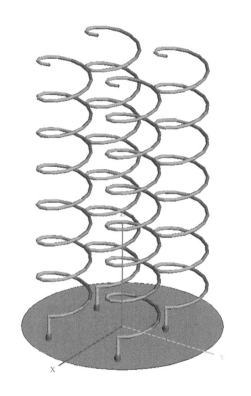

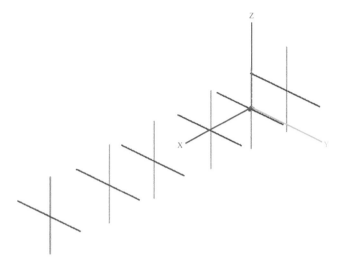

FIGURE P7-3 ▪
A circularly polarized
Yagi-Uda array
antenna with
cross-dipole
elements.

(6) ***Designing a Yagi-Uda Dipole Array.*** Design a six-element Yagi-Uda dipole array antenna using a fixed value for all the directors and their spacings. Compare these results with the optimized arrays presented in Figure 7-28. Explain your observations.

(7) ***Designing a Circularly Polarized Yagi-Uda Array.*** Design a six-element Yagi-Uda array antenna using cross dipole elements. The cross-dipole consists of two wires and two ports, and to achieve circular polarization a phase difference of 90° should be defined between the two ports. The geometrical model of the antenna in FEKO is shown in Figure P7-3.

Frequency-Independent Antennas

8.1 | INTRODUCTION

Broadband communication systems require antennas that can cover a very broad range of frequencies, in some cases with a bandwidth ratio of 40:1. Such a broadband radiation performance cannot be achieved with any of the designs studied in the previous chapters. In this chapter we will study a class of antennas known as frequency-independent antennas.

If an antenna possesses the following characteristics, it is likely to demonstrate a broadband behavior [8]:

- Emphasis on angles rather than lengths
- Self-complementary structures
- Thicker conductors

A frequency-independent antenna should ideally have all these characteristics, but it is still possible for an antenna to demonstrate very wideband behaviors if some of these conditions are met. In addition, a very important feature of a frequency-independent antenna is to have a self-scaling geometry. With this type of configuration, most of the radiation will emanate from a portion of the antenna where it has a resonance length (or circumference) known as the active region. In general, frequency-independent antennas can be categorized into two groups: spiral and log-periodic. In this chapter we will first review the basic properties of these configurations and will then present a few design examples.

8.2 | SPIRAL ANTENNAS

8.2.1 Basics of Spiral Antennas

A spiral antenna is usually constructed to be nearly self-complementary, which results in extremely wideband performance. This type of antenna can be configured in a variety of ways, such as equiangular, Archimedean, rectangular, and conical equiangular. The most basic of these designs, however, is the equiangular planar configuration, and we will briefly review it here.

The equiangular spiral curve is governed by the following generation equation:

$$r = r_0 e^{a\varphi}, \tag{8-1}$$

where r_0 is the radius for $\varphi = 0$, and a is a constant. A positive (negative) sign for a means that the curve is right- (left-) handed. An example of a right-handed equiangular curve is shown in Figure 8-1.

FIGURE 8-1 ▪ An equiangular spiral curve with $r_0 = 1$ and $a = 0.2$.

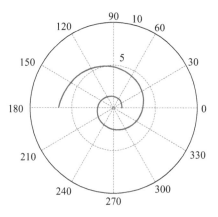

This spiral curve is used to create planar equiangular spiral antennas which are usually constructed using two arms, where each arm is constructed by two spiral curves. Here we show the procedure to design one of the arms for a 1.5-turn equiangular spiral with $a = 0.221$:

$$r_1 = r_0 e^{a\varphi}, \quad r_2 = r_0 e^{a(\varphi - \pi/2)}. \tag{8-2}$$

These two equiangular curves are shown in Figure 8-2a. At $\varphi = 3\pi$, the radius (R) is $8.03r_0$, where r_0 is set to 1 in these figures. The outer radius (R) determines the lower band frequency (i.e., $\lambda_L/4$). The upper band frequency is determined by the radius at the feed point (i.e., $\varphi = 0$), where $r_0 = \lambda_U/4$. This type of design can achieve a bandwidth of 8:1. To truncate the antenna with a smoothly curved, a circle with the outer radius size of R can be defined as shown in Figure 8-2b using dashed line.

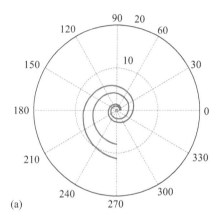

(a)

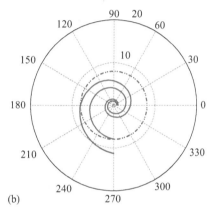

(b)

FIGURE 8-2 ▪ The two equiangular spiral curves for one arm of the antenna: (a) Spiral curves. (b) Spiral curves and truncating boundary.

The final model of this spiral equiangular arm is shown in Figure 8-3. For the outer boundary of the spiral antenna, it is possible to define ellipses instead of circles to achieve a smoother truncation. The adjacent (second) arm can be obtained by rotating this arm by 180°. Readers are encouraged to study the several other available designs for this kind of spiral antenna [7,8]. In the next section we will show the performance of two different types of two-arm spiral antennas.

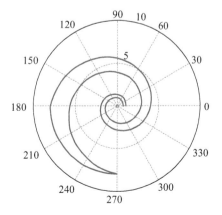

FIGURE 8-3 ▪ One arm of an equiangular spiral antenna.

8.2.2 An Equiangular Spiral Antenna Example

In this section we will study the performance of a planar equiangular spiral antenna designed for the band of 500 MHz to 1 GHz. The geometrical model of this equiangular spiral antenna in FEKO is given in Figure 8-4.

FIGURE 8-4 ▪
Geometry of the
equiangular spiral
antenna in FEKO.

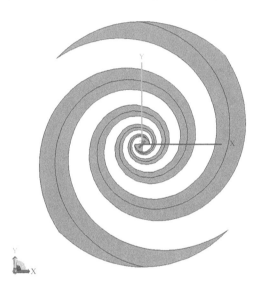

This equiangular spiral antenna has 2.13 turns. The constant a is equal to 0.21. The inner and outer diameters are 4 and 69 cm, respectively. An edge port is defined to feed the antenna. A zoomed view of the edge port is shown in Figure 8-5.

FIGURE 8-5 ▪
Zoomed view of the
feed port in FEKO.

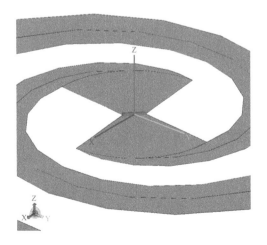

The simulated input impedance of this antenna is given in Figure 8-6. A very smooth variation of the real and imaginary parts of the impedance is observed across the entire band, which is indicative of the frequency-independent features of this design.

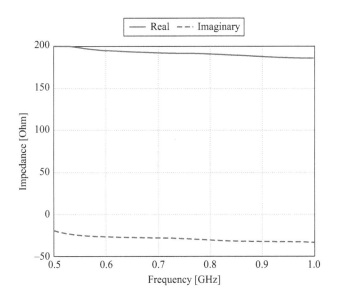

FIGURE 8-6 ▪ Input impedance of an equiangular spiral antenna as a function of frequency.

The gain of this antenna as a function of frequency is shown in Figure 8-7. Note that the variation of gain is about 1.5 dB over the entire band.

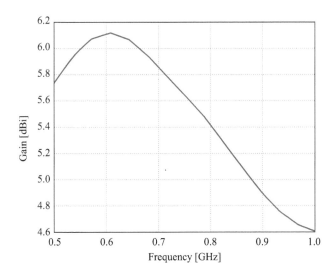

FIGURE 8-7 ▪ Gain of an equiangular spiral antenna as a function of frequency.

The spiral antenna radiation is bidirectional, and the two beams are in the broadside direction. More importantly, the spiral antenna pattern is quite stable over the entire range. The 3D radiation patterns of this spiral antenna across the entire band are shown in Figure 8-8.

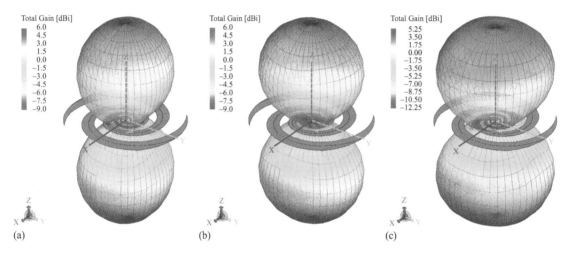

FIGURE 8-8 ▪ 3D radiation pattern of an equiangular spiral antenna: (a) 500 MHz. (b) 750 MHz. (c) 1 GHz.

8.2.3 A Rectangular Spiral Antenna Example

As another example of a two arms spiral antenna, here we will show the performance of a planar rectangular spiral antenna designed for the band of 500 MHz to 2 GHz. The geometrical model of this rectangular spiral antenna in FEKO is given in Figure 8-9.

FIGURE 8-9 ▪
Geometry of the rectangular spiral antenna in FEKO.

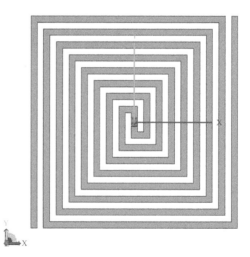

This rectangular spiral antenna has four turns. The inner and outer diameters are 1.8 and 21 cm, respectively. Similar to the equiangular design, an edge port is defined to feed the antenna. The simulated input impedance and gain of this antenna as a function of frequency are illustrated in Figure 8-10 and Figure 8-11, respectively. It is important to point out that in comparison with the equiangular case, the gain for the rectangular spiral increases as frequency increases.

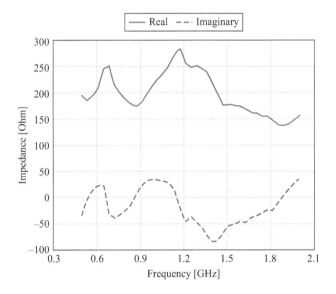

FIGURE 8-10 ■ Input impedance of a rectangular spiral antenna as a function of frequency.

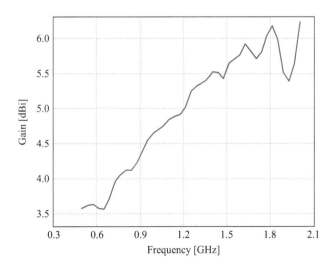

FIGURE 8-11 ■ Gain of a rectangular spiral antenna as a function of frequency.

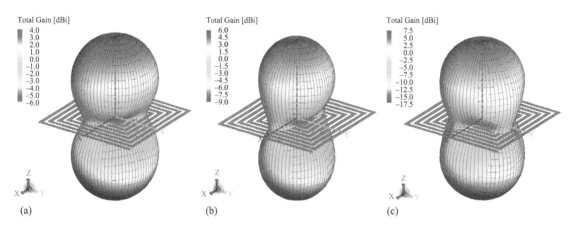

FIGURE 8-12 ■ 3D radiation pattern of an equiangular spiral antenna: (a) 500 MHz. (b) 1.25 GHz. (c) 2 GHz.

As discussed earlier, planar spiral antennas have a bidirectional radiation pattern, which is quite stable across the band. The 3D radiation patterns of this spiral antenna across the entire band are shown in Figure 8-12.

8.3 | LOG-PERIODIC ANTENNAS

8.3.1 Basics of Log-Periodic Antennas

The spiral antennas studied in the previous section clearly demonstrate that introduction of angular dependency in the geometry of the antenna leads to broadband characteristics. On the other hand, constructions of these configurations are quite challenging, so having simpler geometries with straight edges can ease the manufacturing process. One design that can achieve this is a log-periodic structure, where the impedance is repeated periodically as the logarithm of the frequency. Several designs of log-periodic antennas have been developed, such as toothed planar antenna, toothed wedge antenna, toothed trapezoid wedge antenna, zigzag wire antenna, and the log-periodic dipole array (LPDA). The latter is a very practical and simple model of a log-periodic antenna, and a brief review of its design procedure will be outlined here.

A geometrical model of the log-periodic dipole array is given in Figure 8-13. One of the dipoles is directly fed, and the other dipoles are connected to the LPDA transmission line. A virtual wedge, with a subtended angle (α), controls the length of the dipoles.

All the parameters of the LPDA are controlled by the scale factor τ, which is defined as

$$\tau = \frac{R_{n+1}}{R_n} = \frac{L_{n+1}}{L_n}, \tag{8-3}$$

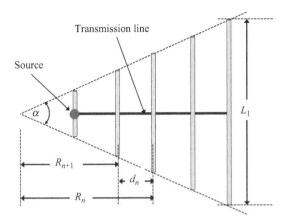

FIGURE 8-13 ◾
The geometry of a
log-periodic dipole
array.

where τ is less than unity. Note that the value of successive element positions and element lengths are equal. The spacing factor of the LPDA is defined as

$$\sigma = \frac{d_n}{2L_n}. \tag{8-4}$$

It can be shown [8] that ultimately this would result in

$$\tau = \frac{R_{n+1}}{R_n} = \frac{L_{n+1}}{L_n} = \frac{d_{n+1}}{d_n}. \tag{8-5}$$

If it is possible to have wires with different radius, then it is desirable to have the radius of the wires also scaled by τ. With such a design, the length of the first dipole (L_1) determines the lower band frequency (i.e., $\lambda_L/2$). Similarly, the length of the last dipole (L_N) determines the upper band frequency (i.e., $\lambda_U/2$). The LPDA broadband antenna has a very simple construction and is also lightweight and low cost, which makes it very popular.

8.3.2 A Log-Periodic Dipole Antenna Example

As an example, here we will show a twelve-element LPDA designed to cover the band from 35 MHz to 60 MHz. For this design the scale factor is 0.93, and σ_0, r_0, and L_0, are set to 0.7, 0.00667, and 2 m, respectively. In this design the wire radius is defined by r_n and is also scaled. For the LPDA transmission line, a nonradiating network connects the ports of all the dipoles. The transmission line has a characteristic impedance of 50 Ω. The voltage source is placed at the center of the shortest dipole, which coincides with the origin of the coordinates system. The schematic model of the network, showing the connections between the ports and transmission lines, is given in Figure 8-14. Twelve ports are defined for the network and are connected together through eleven transmission lines.

The geometry of this dipole antenna modeled in FEKO is shown in Figure 8-15. The feed network connecting the dipole ports together (Figure 8-14)

FIGURE 8-14 ▪
The nonradiating
network that
connects all the
LPDA ports: (a) The
first three ports.
(b) The terminating
elements of the
network.

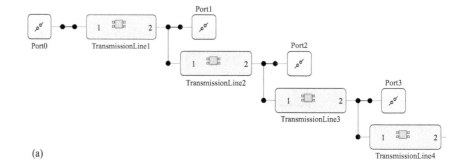

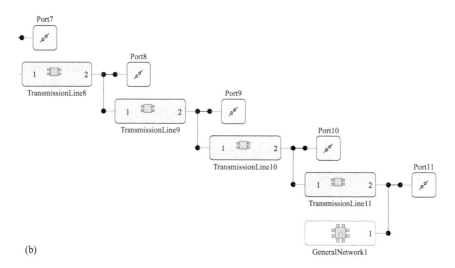

(a)

(b)

FIGURE 8-15 ▪
The geometry of a
log-periodic dipole
array in FEKO.

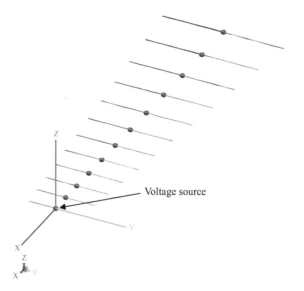

Voltage source

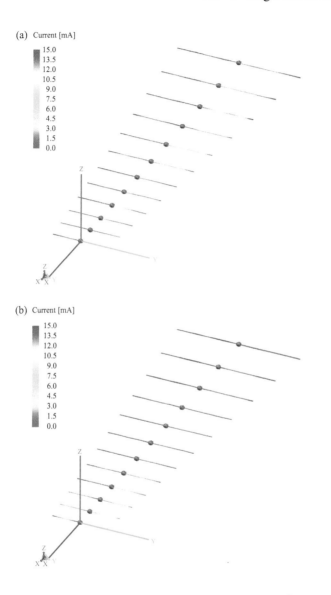

(a) Current [mA]

(b) Current [mA]

FIGURE 8-16 ▪
Current distribution
on the LPDA:
(a) 35 MHz.
(b) 60 MHz.

is nonradiating; therefore, it is not modeled physically here. As discussed earlier, the active region of this antenna changes with frequency. To illustrate this, the current distributions on the dipoles are shown at the extreme frequencies in Figure 8-16. At lower frequency, the current distribution is mainly around the longer end dipoles, and it shifts toward the shorter dipoles at the upper frequency.

The input impedance of this antenna is shown in Figure 8-17. The real part of the impedance is around 50 Ω, and the imaginary part is close to 0 across the entire band. As discussed earlier, this almost constant impedance across the band is indicative of the broadband characteristic of this

FIGURE 8-17 ▪
Input impedance
as a function of
frequency for a
LPDA.

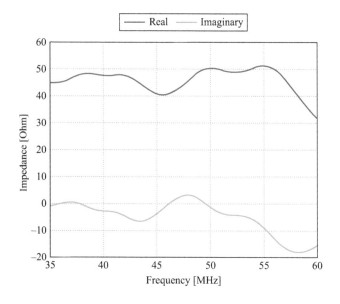

FIGURE 8-18 ▪
Reflection coefficient
at port0 as a function
of frequency for a
LPDA.

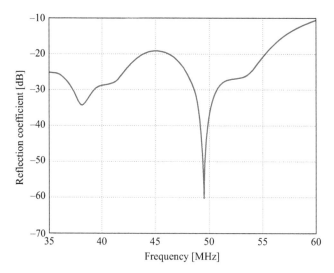

antenna. The fact that the impedance is close to 50 Ω also makes it ideal to be fed with a conventional 50 Ω coaxial cable. The reflection coefficient at the port is presented in Figure 8-18.

The antenna gain as a function of frequency is shown in Figure 8-19. The gain variation is about 1 dB across the entire band. These broadband features clearly show why the LPDA has become such a popular antenna for broadband applications.

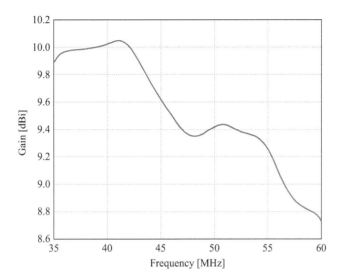

FIGURE 8-19
Gain as a function of frequency for a LPDA.

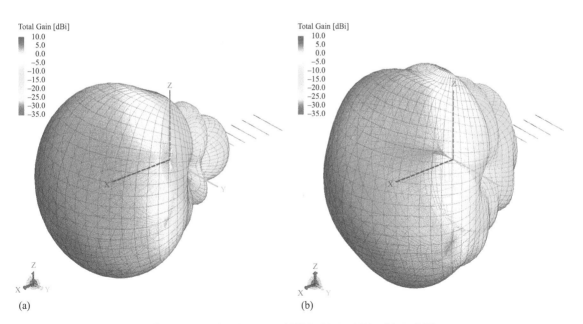

(a)   (b)

FIGURE 8-20 ▪ Current distribution on a LPDA: (a) 35 MHz. (b) 60 MHz.

Similar to a Yagi-Uda antenna, the beam of an LPDA antenna is in the end-fire direction. The 3D radiation pattern of this LPDA is illustrated in Figure 8-20 at the two extreme frequencies. Although some pattern variation is observed in the backlobes, the main beam pattern is quite similar at both frequencies.

EXERCISES

(1) *Cavity-Backed Spiral Antenna.* Place the equiangular spiral cavity designed in section 8.2 on top of a perfect electric cylinder (PEC) cylindrical cavity with a diameter of 75 cm as shown in Figure P8-1 and examine the radiation performance as a function of the cavity height.

(2) *Log-Periodic Dipole.* Following the design procedure given in section 8.3, design an LPDA to cover the frequency band of 54 to 216 MHz. This is the frequency band of VHF-TV and FM broadcast. The bandwidth ratio for this design is 4:1.

FIGURE P8-1 ■
A cavity-backed spiral antenna.

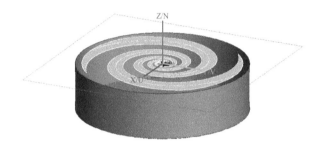

Horn Antennas

Chapter Outline

9.1 | INTRODUCTION

Horn antennas are very popular in microwave applications for frequencies above 1 GHz. They provide high gain, low return loss, and relatively wide bandwidth and are rather easy to construct. The horn is widely used as a feed element for large radio astronomy, satellite tracking, and communication dishes installed throughout the world. In addition to its utility as a feed for reflectors and lenses, it is a common element of phased arrays and also serves as a universal standard for calibration and gain measurements of other high-gain antennas.

While several different types of electromagnetic horns are available, the most conventional types are the pyramidal and conical horn antennas. In this chapter, we will first study the performance of sectoral horn antennas to clearly demonstrate the advantage of waveguide flare opening. Next, pyramidal horn antennas are examined in detail. Finally, some other practical designs such as the conical horn antenna, the multimode Potter horn antenna, and the corrugated conical horn antenna will be demonstrated.

9.2 | SECTORAL HORN ANTENNAS

A horn antenna is analogous to a megaphone; that is, the structure is flared such that it provides directivity for the waves. It is usually fed by a waveguide, and the flaring acts as a transition from the waveguide mode to the free-space mode. In sectoral horns the waveguide is flared in only one direction. If the broad wall of the waveguide is flared, this structure is known as an E-plane sectoral horn. In other words, the waveguide is flared in the direction of the E-field. Alternatively, the H-plane sectoral horn is flared in the narrow wall of the waveguide and the broadside is left unchanged. A geometrical view of E- and H-plane sectoral horn antennas is given in Figure 9-1.

FIGURE 9-1 ▪
Geometrical model of sectoral horn antennas in FEKO: (a) E-plane. (b) H-plane.

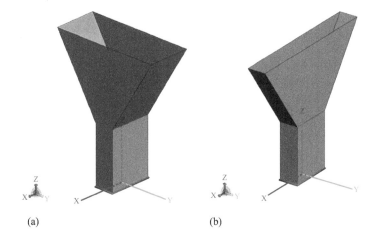

(a)   (b)

To generate the physical model of a sectoral (or pyramidal) horn antenna in FEKO, we need to create two geometries: (1) the waveguide section, which can easily be generated by defining cuboids; (2) and the flared section of the horn, which can be created using a flare. The user interfaces for creating these two geometries with FEKO are presented in Figure 9-2. Once the dimensions of the horn are known, creating the geometry is quite straightforward. After these two geometries are created, they should be united together, and the appropriate faces on this solid (opening of the waveguide and the horn aperture) must be deleted to complete the geometrical model of the waveguide-fed horn antenna.

To excite the antenna, we first define a waveguide port. The waveguide port is the end of the waveguide section in the model as shown in Figure 9-3a. Once the port is defined, we assign a waveguide excitation and excite only the

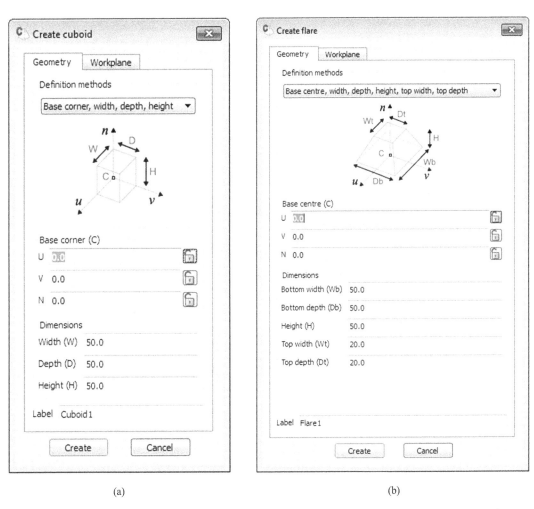

(a)   (b)

FIGURE 9-2 ▪ Objects for creating a horn antenna with FEKO: (a) Cuboid for waveguide section. (b) Flare for the flared section of the horn. The dimensions in these figures are FEKO default setup.

fundamental mode (Figure 9-3b). The dimensions of the waveguide should be set to ensure that the required modes can propagate.

To see the radiation performance of these configurations, we will study two X-band sectoral horn antennas. The standard dimension of an X-band waveguide (i.e., WR-90 waveguide) is 9.0×4.0 in (22.86×10.16 mm). For both configurations the height of the flare is 3.82 cm. For the E- and H-plane sectoral horns, the flare width is 5.72 and 7.74 cm, respectively. These dimensions are selected from a pyramidal horn antenna that will be studied later on. The radiation patterns of these two sectoral horn antennas are given in Figure 9-4.

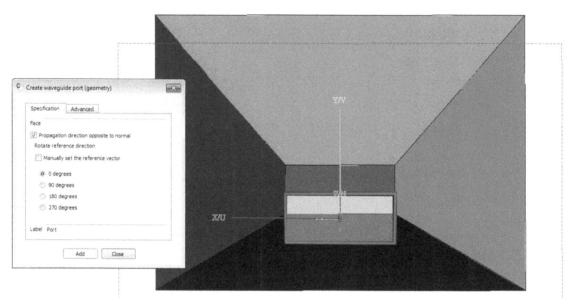

(a)

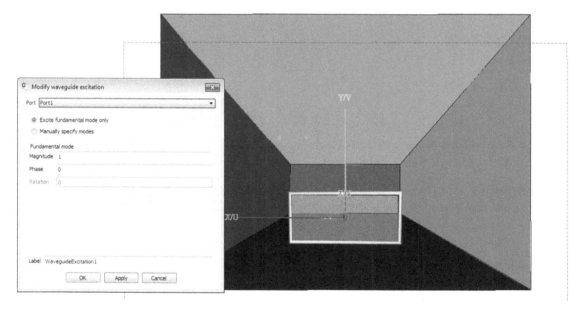

(b)

FIGURE 9-3 ▪ Excitation of the horn antenna: (a) Waveguide port. (b) Waveguide excitation.

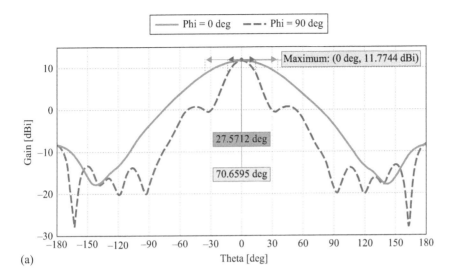

(a)

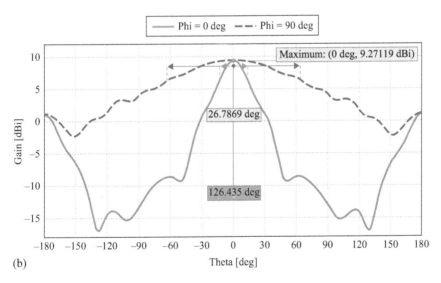

(b)

FIGURE 9-4 ▪
Radiation patterns
of sectoral horn
antennas: (a) E-plane.
(b) H-plane.

As expected, for both designs the beam is narrower in the plane that has been flared. In addition, the E-plane sectoral horn shows a higher sidelobe level [7]. The 3D radiation pattern for these horn antennas is presented in Figure 9-5.

9.3 | PYRAMIDAL HORN ANTENNAS

The most common type of horn antenna is the pyramidal horn, which is flared in both directions. Its radiation characteristics are essentially a combination of

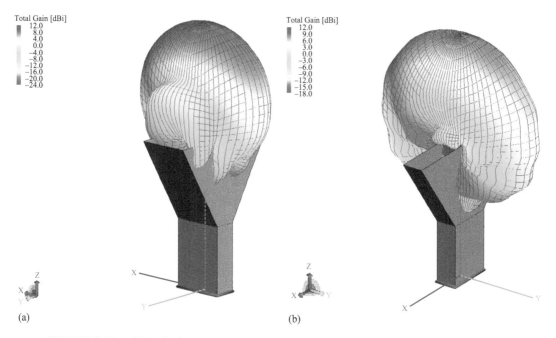

FIGURE 9-5 ▪ 3D radiation patterns of sectoral horn antennas: (a) E-plane. (b) H-plane.

the E- and H-plane sectoral horns. A cross sectional view of the schematic model of the pyramidal horn is shown in Figure 9-6.

The equivalence principle is used to obtain the electromagnetic fields on the aperture of the horn antenna. The radiation pattern of the antenna is then determined by far-field transformation of the equivalent tangential currents. For the fundamental mode of a rectangular waveguide, the electric and magnetic

FIGURE 9-6 ▪
Cross section of a
pyramidal horn:
(a) y-z plane.
(b) x-z plane.

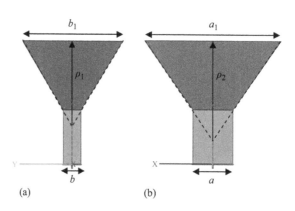

current densities have a cosinusoidal amplitude distribution along the broad direction of the aperture and a quadratic phase variation in both the orthogonal directions. Mathematically, this is given by

$$E'_y(x', y') = E_0 \cos\left(\frac{\pi}{a_1} x'\right) e^{-j[k(x'^2/\rho_2 + y'^2/\rho_1)/2]}, \tag{9-1}$$

$$H'_x(x', y') = -\frac{E_0}{\eta} \cos\left(\frac{\pi}{a_1} x'\right) e^{-j[k(x'^2/\rho_2 + y'^2/\rho_1)/2]}. \tag{9-2}$$

The far-field radiation pattern is then computed directly from these aperture fields. Detailed formulations for the aperture field distributions and the radiation patterns of the pyramidal horn are available in the literature [7]. Here we will summarize the final equations. The far-field electric fields for a pyramidal horn antenna are

$$E_\theta = j\frac{kE_0 e^{-jkr}}{4\pi r} [\sin\phi(1 + \cos\theta)I_1 I_2], \tag{9-3}$$

$$E_\phi = j\frac{kE_0 e^{-jkr}}{4\pi r} [\cos\phi(1 + \cos\theta)I_1 I_2]. \tag{9-4}$$

where I_1 and I_2 are given by

$$I_1 = \int_{-a_1/2}^{+a_1/2} \cos\left(\frac{\pi}{a} x'\right) e^{-jk[(x'^2/2\rho_1) - x'\sin\theta\cos\phi]} dx', \tag{9-5}$$

$$I_2 = \int_{-b_1/2}^{+b_1/2} e^{-jk[(y'^2/2\rho_1) - y'\sin\theta\sin\phi]} dy'. \tag{9-6}$$

It is important to note that the solution to these two integrals is in the form of Fresnel integrals.

The desired gain and the dimensions of the rectangular feed waveguide must be known to design a pyramidal horn antenna. Here we used the procedure given in [7] to design an X-band pyramidal horn antenna for 15 dBi gain. The waveguide feed is the WR-90 standard used in the previous section. The height of the flare is 3.82 cm, and the aperture of the horn is 5.72×7.74 cm. The geometrical model of this pyramidal horn antenna in FEKO is given in Figure 9-7. In this configuration, the fundamental mode of the waveguide means that the electric fields are in the y direction. The amplitude and phase of the y component of the electric field on the horn aperture are shown in Figure 9-8. As expected, the fields are maximum at the center. The phase distribution on the aperture also indicates that the rays seem to be emanating from a point inside the horn. This point is known as the phase center of the horn antenna [7].

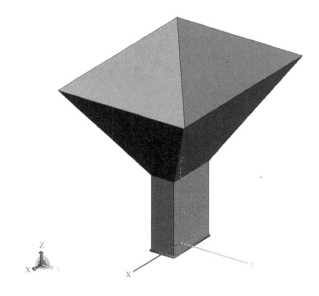

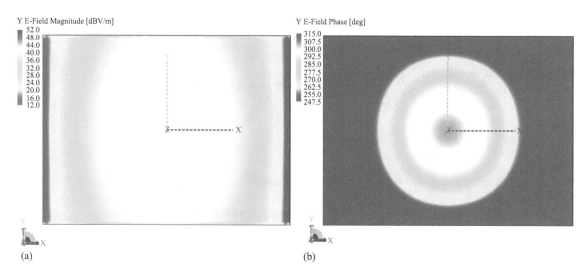

(a)　　　　　　　　　　　　　　　　　　　　(b)

FIGURE 9-8 ■ Electric fields on the aperture of a horn antenna: (a) $|E_y|$ in dB. (b) Phase of E_y in degrees.

The radiation pattern of this pyramidal horn antenna is given in Figure 9-9. This design exactly achieves the desired value of 15 dB gain, and E-plane ($\varphi = 90°$) has a narrower beam and also a higher sidelobe. In general, pyramidal horn antennas cannot achieve a symmetric pattern, but they are still widely used as standards to make gain measurements of other antennas. As such, they are also known as standard gain horns.

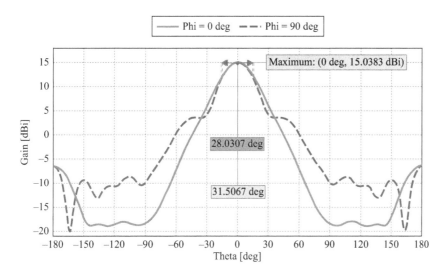

(a)

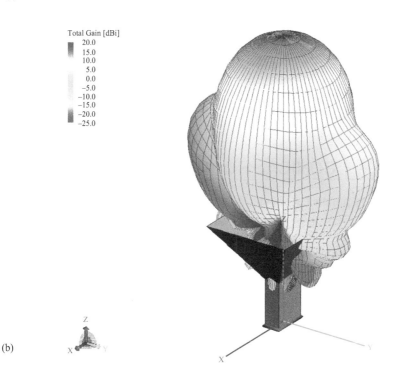

(b)

FIGURE 9-9 ■
Radiation pattern of an X-band pyramidal horn antenna in FEKO: (a) patterns in the principal planes. (b) 3D pattern.

It is also worth pointing out that theoretical analysis method described earlier can provide reasonably good agreement with these full-wave simulation results. Figure 9-10 compares this analytical solution [equations (9-3 and 9-4)] with the results obtained by FEKO.

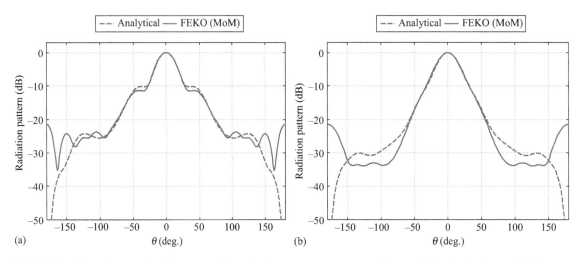

FIGURE 9-10 ■ Comparison between the radiation patterns computed by analytical solution and full-wave simulation for an X-band pyramidal horn antenna: (a) E-plane. (b) H-plane.

9.4 | CONICAL HORN ANTENNAS

Another very practical horn configuration is the conical horn antenna. It is fed by a circular waveguide, in contrast to the sectoral and pyramidal horns, which are fed by rectangular waveguides. Similar to the pyramidal horn antenna studied in the previous section, the geometrical model of the conical horn is created by defining two objects: a cylindrical waveguide section and a cone. The user interfaces for creating these two geometries with FEKO are shown in Figure 9-11.

A conical horn antenna's analysis process and general behavior are similar to the pyramidal horn. As the flare angle increases, the directivity for a given horn length increases until it reaches a maximum. Beyond that point the gain will decrease. Here we will show only the performance of a 32 GHz Ka-band conical horn antenna designed to achieve a gain of 15 dBi. The radius of the waveguide is 0.323λ. The cone has a height of 1.44λ and an aperture radius of 1.04λ. The horn is fed with a pin that will excite the fundamental mode of the circular waveguide (i.e. the TE_{11} mode). This pin has a length of $\lambda/4$, and is

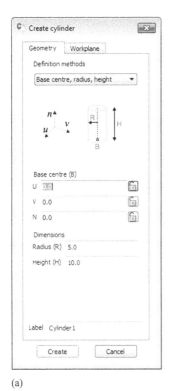

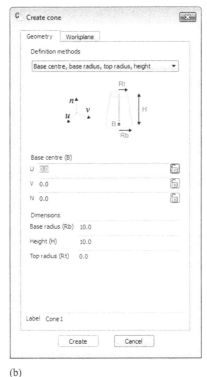

(a)   (b)

FIGURE 9-11 ▪
Objects for creating a conical horn antenna with FEKO: (a) Cylinder for waveguide section. (b) Cone for the flared section of the horn.

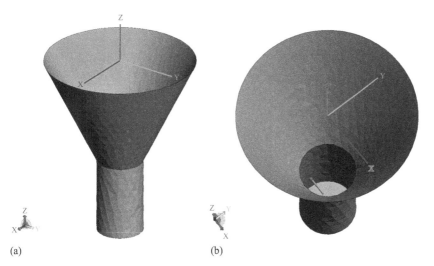

(a)   (b)

FIGURE 9-12 ▪
A Ka-band conical horn antenna modeled in FEKO.

place at a distance of $\lambda/4$ from the end of the circular waveguide. The geometrical model of this conical horn antenna in FEKO is given in Figure 9-12.

The total electric field on the aperture of the horn is shown in Figure 9-13. The maximum field strength is observed along the x direction at $y = 0$.

FIGURE 9-13 ■
Magnitude of the
total electric fields
on the aperture of a
horn antenna.

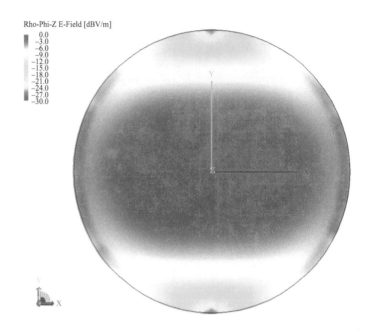

FIGURE 9-14 ■ Radiation pattern of an Ka-band conical horn antenna in FEKO: (a) patterns in the principal
planes, (b) 3D pattern.

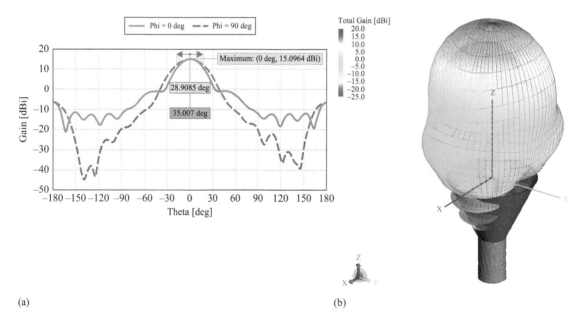

The radiation pattern of this pyramidal horn antenna is given in Figure 9-14. The design also achieves the desired value of 15 dBi gain. Similar to the pyramidal horn example in the previous section, the E-plane ($\varphi = 0°$) has a narrower beam and a higher sidelobe. Also, these antennas cannot achieve a symmetric pattern with the fundamental mode excitation. However, azimuthally symmetric patterns with the proper excitation of waveguide modes are possible with the circular geometry of these horn antennas.

9.5 | A MULTIMODE HORN ANTENNA: THE POTTER HORN

In many applications such as reflector feeds, a horn antenna with a symmetric radiation pattern is desired; however, conventional pyramidal and conical horn antennas cannot achieve such radiation performance. One of the most fundamental methods of achieving such a performance is to excite higher-order modes in the horn waveguide. Here we will briefly discuss the basics of higher order mode generation and demonstrate a Potter horn antenna design procedure [20,21] that was used to achieve a symmetric radiation pattern with specific requirements.

Conical horn antennas use the dominant TE_{11} mode in the circular waveguide and generate a directive beam with an asymmetric radiation pattern. Unity azimuthal modes of TE_{1m} or TM_{1m} ($m > 1$) can be excited by abrupt or gradual changes in the diameter of the horn or waveguide. This change in diameter will not excite TE_{nm} or TM_{nm} ($n \neq 1$). The easiest method of exciting higher-order modes with unity azimuthal dependence is to introduce a step change into the diameter of the horn. The abrupt junction will force a break in the smooth current pattern, and if the normalized output radius is greater than the cutoff wavenumber for the desired mode then some power will be transferred to that model. The amount of power will depend on both radii. In addition, changes in the flare angle of a horn will also excite higher-order modes.

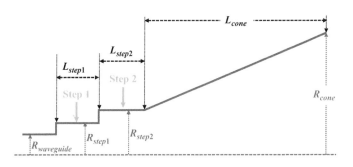

FIGURE 9-15 ◾
Cross sectional view of the Potter horn antenna.

To achieve a symmetric pattern, a common requirement is to excite a proportion of TM_{11} mode to add to the TE_{11} mode, which is done in the Potter horn antenna. A cross sectional geometry of the Potter horn antenna is given in Figure 9-15.

While several parameters have to be tuned in a design to achieve the desired radiation pattern, some fundamental design rules can be given here. First, the output radius should be greater than 0.6098λ such that the TM_{11} mode can propagate; otherwise this mode will be evanescent. In addition, the output radius should not be greater than 0.8485λ to avoid the propagation of the TE_{12} mode. It is also preferable to have the desired amount of TM_{11} power excited (in the flared section of the horn) before this radius is reached. The amount of power that is normally required to be transferred to the TM_{11} mode is between 9 and 20%.

For a practical design, an inner radius greater than 0.53λ is required. Since this value is greater than the radius of the single mode circular waveguides, two steps are needed for a Potter horn. The first junction keeps the power in the TE_{11} mode, and the second junction excites the TM_{11} mode. While in many cases an optimization would be required for the horn design, basically the remainder of the horn design process is to flare the horn so that the aperture diameter will radiate the desired beamwidth. The main challenge in the design is that the TM_{11} mode must be brought in phase with the TE_{11} mode while controlling the other constraints.

As with reflector feeds, typically a symmetric radiation pattern with specific requirements such as half-power beamwidth, or a $\cos^q(\theta)$ radiation pattern model, is required. Here we present a design for a Potter horn antenna that achieves a symmetric radiation pattern with $q = 6.5$ at the center design frequency.

For any design, the initial dimensions of the horn parameters are determined using the design guidelines presented in the this section; however, in most cases it is necessary to tune these dimensions to achieve the desirable pattern. The particle swarm optimizer (PSO) in FEKO was selected for this task. In total, seven parameters have to be optimized for this design: the radius of the waveguide feed ($R_{waveguide}$); the radius and length of two waveguide steps (R_{step1}, L_{step1}, R_{step2}, L_{step2}); and the radius and length of the cone (R_{cone}, L_{cone}). In FEKO optimizations are done using the OPTFEKO solver. Readers are referred to the FEKO manual for details on the available optimization tools and setups. For this design, at each fitness evaluation during the optimization, the radiation pattern is computed at a number of discrete points chosen to match the required $\cos^q(\theta)$ pattern and to achieve a symmetric pattern in the two principal planes. The optimized dimensions of the Potter horn are given in Table 9-1.

TABLE 9-1 ▪ Optimized dimensions of a Potter horn antenna.

$R_{waveguide}$	R_{step1}	L_{step1}	R_{step2}	L_{step2}	R_{cone}	L_{cone}
0.323λ	0.571λ	0.386λ	0.763λ	1.539λ	1.009λ	0.848λ

The total electric fields inside the optimized Potter horn antenna, at different sections as described in Figure 9-15, are shown in Figure 9-16. The formation of the TM_{11} mode in the Potter horn can clearly be seen in Figure 9-16c.

(a)   (b)   (c)   (d)

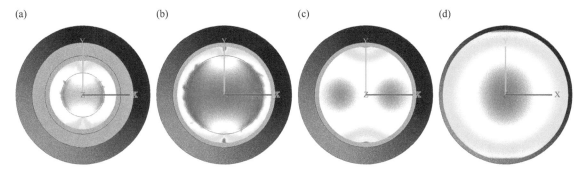

FIGURE 9-16 ▣ Electric field magnitude inside the optimized Potter horn antenna: (a) Start of step 1. (b) Start of step 2. (c) End of step 2. (d) Horn aperture.

Symmetric aperture fields are generated at the horn aperture, which should lead to a symmetric radiation pattern. The optimized Potter horn and the 3D gain pattern of the antenna are visible in Figure 9-17. The normalized radiation pattern of the optimized Potter horn and a comparison with the ideal cosine q model are given in Figure 9-18.

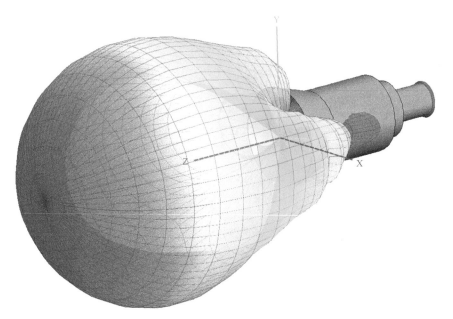

FIGURE 9-17 ▣ Radiation pattern of an optimized Potter horn antenna at 32 GHz.

FIGURE 9-18 ◾
Normalized radiation
patterns of an
optimized Potter
horn antenna.

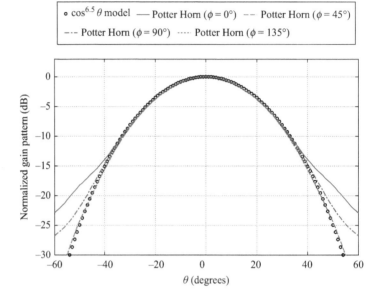

The radiation pattern of the optimized Potter horn is almost completely symmetric in the four plane cuts and matches quite well with the ideal cosine q model up to 40°. Although the beam width requirement for a feed horn design depends on the system parameters, in most cases 40° is sufficient for reflector feeds. It is important to point out that typically multimode horns such as the Potter design have a narrow bandwidth, but this should be weighed against the simplicity of the design compared with a corrugated horn.

9.6 | CORRUGATED CONICAL HORN ANTENNA

The need to reduce spillover and cross-polarization losses, in addition to improving aperture efficiencies of large reflectors, inspired a large emphasis on horn antenna research. For many years, smooth-wall circular and rectangular horns were the only known types, but the benefits of corrugating surfaces or walls became known in the 1940s and 1950s.

Corrugations are necessary to support the hybrid HE_{11} mode, which is the most desirable horn radiation performance. To that end, the inner surface of the horn must be anisotropic in such a way that it has different reactances in the azimuthal and the propagation direction. With these conditions the tranverse electric (TE) and transverse magnetic (TM) components become locked together as a single hybrid mode and propagate with a unique common velocity. With a proper design, the balanced hybrid condition is approximated over a wide range of frequencies up to about 2.2:1. Operation is limited at the upper end of the band by the appearance of the unwanted EH_{11} mode.

The radiation from a conical corrugated horn operating in the HE_{11} dominant mode near the balanced condition can be obtained from the aperture fields as discussed in [22]. In practice, however, these assumptions are valid only up to $\theta = 35°$. An accurate analysis requires a full-wave simulation, as will be outlined here.

A waveguide-fed corrugated conical horn is designed using the commercial software Antenna Magus [23] for the operating frequency of 10 GHz and 15 dBi gain. The geometry of the horn model and the parameters are given in Figure 9-19. The radius of the horn aperture is 32.77 mm.

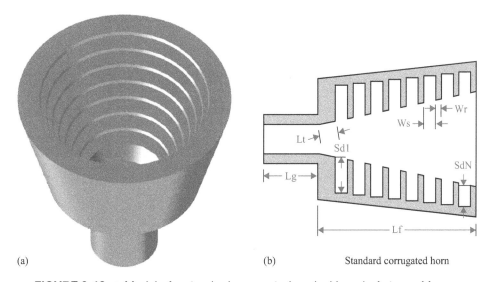

(a) (b) Standard corrugated horn

FIGURE 9-19 ▪ Model of a standard corrugated conical horn in Antenna Magus.

Once the antenna is designed in Antenna Magus, it is exported into FEKO. That is, a *.cfx file is generated and can be opened directly with FEKO. The geometry of the antenna in FEKO along with the surface currents on the horn antenna is given in Figure 9-20.

As discussed earlier, a notable advantage of the corrugated conical horn is the symmetric radiation pattern of the antenna. The radiation patterns of the antenna at 10 GHz are given in Figure 9-21. The E- and H-plane patterns match very well up to about 20 dB down from the maximum gain.

FIGURE 9-20 ▪
Corrugated conical
horn in FEKO.

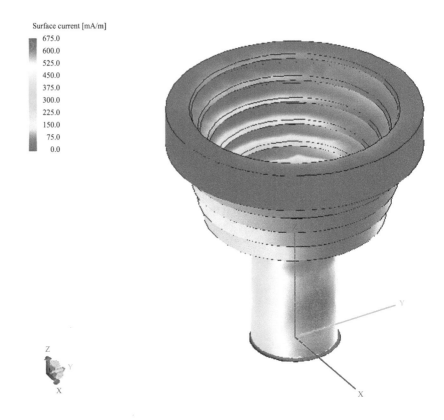

FIGURE 9-21 ▪
Radiation patterns of
a corrugated conical
horn in FEKO:
(a) patterns in the
principal planes,
(b) 3D pattern.

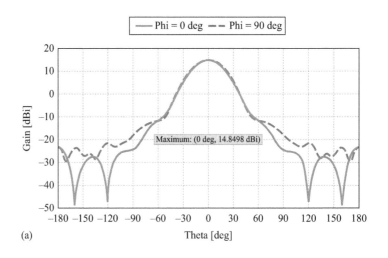

(a)

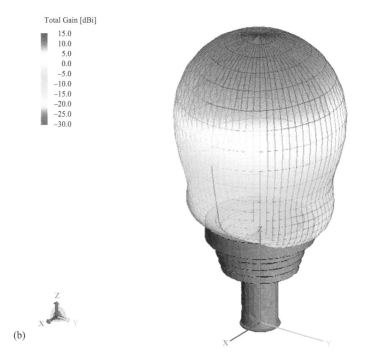

Total Gain [dBi]

FIGURE 9-21 ■
(*continued*).

(b)

EXERCISES

(1) *Sectoral Horn Antennas.* For the E-plane sectoral horn antenna given in section 9.2, study the effect of flare angle on the performance of the horn. Start this study from the case of the open-ended waveguide (i.e., when flare angle is zero) until you reach a flare angle of 60°. At what flare angle does the horn antenna achieve the maximum gain? Repeat this problem for the H-plane sectoral horn.

(2) *Pyramidal Horn Antenna.* Design three S-band pyramidal horn antennas to achieve gains equal to 9, 15, and 20 dBi. Use standard dimension for the S-band waveguide. Compare the radiation patterns with the analytical solutions given in [7].

(3) *Conical Horn Antenna.* Design an X-band optimum directivity conical horn antenna with a cone length of 10λ. What is the directivity of this horn antenna?

Reflector Antennas

Chapter Outline

10.1 | INTRODUCTION

High-gain antennas are an essential part of long-distance radio communication links and high resolution radar systems. Reflector antennas have been in use since Heinrich Hertz's discovery of electromagnetic wave propagation. Reflector systems are the most widely used high-gain antennas and can achieve gains far in excess of 30 dB in the microwave region, which would be extremely difficult (if not impossible) with any other single antenna we have studied so far. Over the years, different theoretical techniques have been developed for analysis of reflector antennas. In this chapter we will discuss some basic design guidelines as well as examples of corner, parabolic, and spherical reflector antennas. More emphasis is given to the parabolic reflector antenna since this is the most practical reflector antenna configuration in use today.

10.2 | CORNER REFLECTOR ANTENNAS

A simple flat perfect electric conductor (PEC) ground plane can be used to direct energy in a desired direction, such as using a dipole antenna placed above a ground plane (Chapter 2). With an electrically large ground plane that can be approximated as infinite surface, image theory can be used to analyze the radiation characteristics of these systems. If the dipole antenna is properly placed, the antenna gain can be increased in these configurations (Chapter 2). However, a flat PEC ground plane does not properly collimate the beam.

For better collimation of energy, the shape of the reflector must be changed. One of the simplest arrangements is two plane reflectors joined to form a corner. A geometrical model of the corner reflector is given in Figure 10-1. Because of its simplicity, it has many unique applications.

FIGURE 10-1 ▪
Geometrical model of a corner reflector.

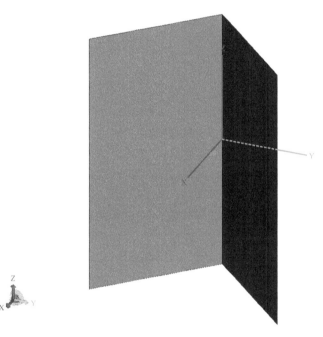

In most cases, the subtended angle formed by the corner reflector plates is usually 90°. The spacing between the vertex and the feed element is the most important parameter in realizing good system efficiency, and it typically increases as the subtended angle of the reflector decreases. In the theoretical analysis of corner reflectors, it is typically assumed that the plates are infinite in extent; however, in practice the dimensions are finite. Similarly, the feed element is assumed to be a line source, but in practice the feed for corner reflectors is usually a dipole antenna.

The general guidelines for corner reflector antennas are given in [7], and readers are encouraged to become familiar with these rules of thumb for the design. Here we will demonstrate the performance of corner reflector antennas with the following parameters:

- Reflector height $= 2\lambda$
- Aperture width $= 1.5\lambda$

- Feed-to-vertex distance = 0.5λ
- Dipole length = 0.475λ

The length of each plate is then determined by the subtended angle between the two plates. The geometrical model of a corner reflector with a 90° angle and an ultra high frequency (UHF) dipole feed element is given in Figure 10-2.

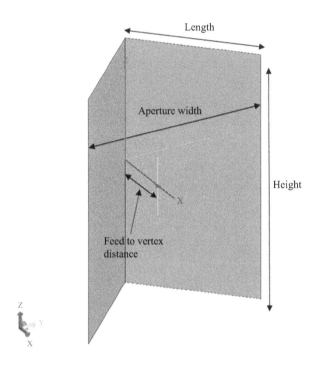

Length

Aperture width

Height

Feed to vertex distance

FIGURE 10-2 ■
A 90° corner reflector antenna with a dipole feed.

At the center frequency of 300 MHz, the wavelength is approximately 1 m. Thus, the corner reflector has a height of 2 m and an aperture opening of 1.5 m.

To efficiently analyze large reflectors such as the corner reflector studied here, one approach is to use the physical optics (PO) solver in FEKO for the reflector and the method of moments (MoM) solver for the feed element. To assign the PO solver for the analysis of the reflector, the reflector face is selected and the PO solution is assigned. The face properties of the corner reflector and the solution setup in FEKO are shown in Figure 10-3.

More discussion on the solver selections will be given when we study the parabolic reflector antenna. For the 90° corner reflector antenna in this section, the current distribution on the plates and the 3D radiation pattern are shown in Figure 10-4.

FIGURE 10-3 ■
Solution setup for a
reflector in FEKO.

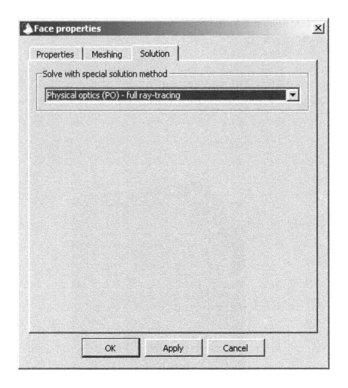

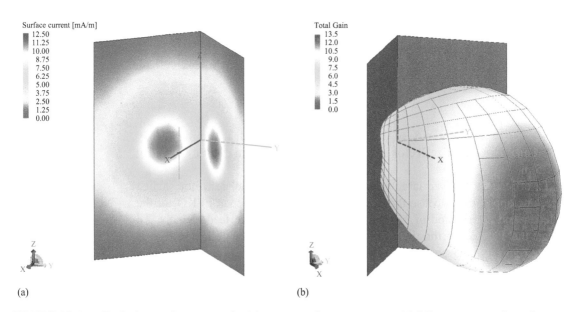

(a)   (b)

FIGURE 10-4 ■ Radiation performance of a 90° corner reflector antenna: (a) PO currents on the reflector.
(b) 3D radiation pattern of the antenna at 300 MHz.

Compared with the flat ground planes studied earlier, the beam of the corner reflector antenna is pointing in the forward direction. To also observe the effect of subtended angle between the two plates (Figure 10-5), the radiation patterns of a dipole element in isolation, in front of a flat pate reflector, and in between corner reflectors with different subtended angles are shown in Figure 10-6.

FIGURE 10-5 ■ 2D cross sectional view of a corner reflector in FEKO.

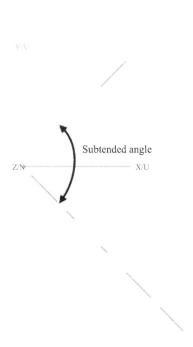

These results show that the flat reflector has the lowest gain in the forward direction. The 90° corner reflector, on the other hand, achieves the highest gain. For the corner reflectors, as the subtended angle increases the forward gain of the reflector decreases. Again, the feed position was held constant at 0.5λ in all these cases. This position is optimum for the 90° corner reflector but not for the other corner reflectors studied in this section.

10.3 | DESIGN AND ANALYSIS OF PARABOLIC REFLECTOR ANTENNAS

10.3.1 Basic Principles

In this section we will study the performance of the popular parabolic reflector antenna. Even though a corner reflector can collimate the beam better than a

FIGURE 10-6 ▪
Radiation patterns at
300 MHz for various
subtended angles of
the corner reflector.

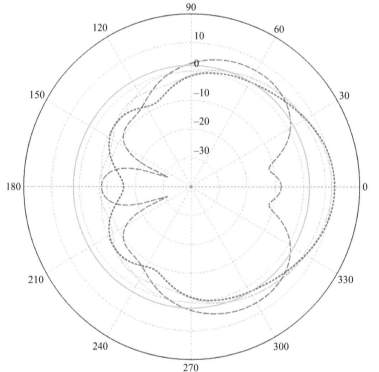

planar reflector surface it typically is not ideal for achieving high aperture efficiency. The parallel rays that are incident on the reflector will converge at the focal point in an ideal collimating system, which is achieved by a parabolic reflector surface [7,8].

The parabolic reflector antenna, also known as dish, is a paraboloid of revolution. The geometrical equations for a parabola are well known, so here we will focus only on considerations in designing the reflector antenna. The most important parameters in reflector designs are the diameter of the dish (D) and the focal length (F). For axisymmetric reflectors, these two parameters are sufficient to completely specify the configuration. In addition, reflectors are often specified by the ratio of focal length to diameter (i.e., F/D), where the typical values are around 1.

The reflector is an aperture type antenna, and as such the gain is directly proportional to the size of the aperture (i.e., D). The main challenge in reflector design is typically matching the feed antenna pattern to the reflector, which is done by selecting the proper focal length for the design. Usually the goal is to have a taper at the edge that is about 10 dB down from the reflector apex. Consequently, proper design of the feed antenna is of significant importance in designing the reflector antenna. For the feed antenna it is usually preferable to have an element that provides an azimuthally symmetric pattern. As such, in most cases corrugated conical horn antennas are used as reflector feeds. Importantly, the feed antenna must also be positioned such that the phase center of the feed is placed at the focal point. For horn antennas the phase center is usually between its aperture and imaginary apex point [7].

10.3.2 Axisymmetric Parabolic Reflectors

Here we will study the performance of an X-band axisymmetric parabolic reflector antenna designed for a nominal gain of 35 dBi. The feed antenna is the corrugated conical horn antenna studied in Chapter 9. The parabolic reflector is a perfect collimating device and is therefore frequency independent; however, the bandwidth of the reflector is typically controlled by the feed antenna.

As noted earlier, in aperture type antennas such as reflectors the antenna gain is proportional to the electrical size of the aperture. Therefore, if the desired gain for the antenna is specified, the required aperture size can be determined. The maximum directivity of an aperture with uniform illumination is given by

$$D = 4\pi \frac{A}{\lambda^2},$$
(10-1)

where A is the size of the aperture. The gain of the antenna is then determined by taking into account the aperture efficiency, that is,

$$G = \eta D,$$
(10-2)

where η is the aperture efficiency. The dominant terms in aperture efficiency for reflector antennas are typically illumination and spillover efficiencies. Illumination efficiency is a measure of how well the entire aperture of the reflector is being illuminated. For aperture antennas, the maximum directivity is obtained when the aperture has a uniform illumination. Spillover efficiency is a measure of the amount of radiated feed antenna power intercepted and collimated by the reflector aperture. More details on efficiency analysis for reflector antennas can be found in [7,8].

One rather simple and practical approach to design an axisymmetric reflector is to control the edge taper of the system. To better illustrate this design procedure, first let us review the geometrical model of the reflector. A schematic cross sectional view of an axisymmetric reflector along with the design parameters is given in Figure 10-7. As pointed out earlier, all the parameters can be determined if F and D are specified.

FIGURE 10-7 ■
Cross sectional view
of an axisymmetric
parabolic reflector.

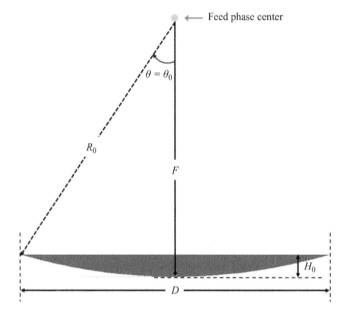

The reflector height (H_0) and the subtended half-angle (θ_0) are calculated using

$$H_0 = \frac{D^2}{16F}, \qquad (10\text{-}3)$$

$$\theta_0 = 2 \tan^{-1}\left(\frac{1}{4(F/D)}\right). \qquad (10\text{-}4)$$

The distance from feed phase center to reflector rim (R_0) is given by

$$R_0 = \frac{F - H_0}{\cos\theta_0}. \qquad (10\text{-}5)$$

Now that these geometrical parameters are defined, let us study the edge taper. Edge taper is the ratio of feed power at the edge (rim) of the reflector to that at the center. Mathematically this is given by

$$ET = 20 \log_{10}\left(\frac{E_f(\theta = \theta_0)/R_0}{E_f(\theta = 0)/F}\right), \qquad (10\text{-}6)$$

where E_f is the radiation pattern of the feed antenna. A typical model for this function is

$$E_f(\theta) = \cos^q \theta. \tag{10-7}$$

As discussed earlier, it is desirable to have an edge taper in the order of -10 dB. This will ensure a good compromise between illumination and spillover efficiencies. Depending on the feed horn, the value of q would be different, but the typical values are between 5 and 10.

Now let's go back to designing the parabolic reflector. To achieve a gain of 35 dBi, the size of the dish is selected to be 20 wavelengths. The maximum aperture directivity for an aperture this size is about 36 dBi. Based on efficiency and edge taper considerations, an F/D of 0.9 is selected for this design. The cross sectional view of this parabolic reflector is presented in Figure 10-8.

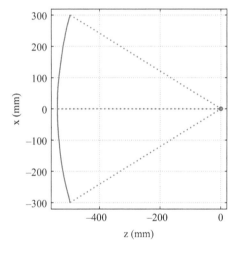

FIGURE 10-8 ■
Cross sectional view of an X-band axisymmetric parabolic reflector antenna.

The geometrical model of the reflector antenna in FEKO is shown in Figure 10-9.

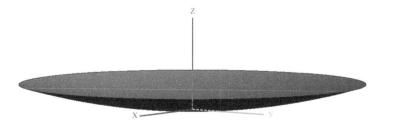

FIGURE 10-9 ■
An axisymmetric parabolic reflector antenna modeled in FEKO.

Four different analysis techniques are introduced here to analyze the performance of this parabolic reflector, which is categorized into two groups. In the first group, the radiation pattern of the horn antenna is used as the excitation.

In FEKO, this can be done by simulating the horn antenna separately and generating an *.ffe file, which is then used to excite the reflector. In the second group the full system (i.e., both the reflector and horn antenna) is modeled. The full reflector system model is illustrated in Figure 10-10.

FIGURE 10-10 ▪
An axisymmetric parabolic reflector antenna and feed horn modeled in FEKO.

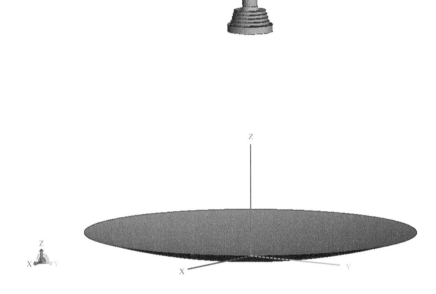

The primary computational challenge in analyzing reflector antennas is their large electrical size. For efficient analysis, two options are available: physical optics and multilevel fast multipole method (MLFMM). If a point source radiation pattern is used for the feed, only the solver for the reflector must be specified. If the complete system is modeled, one may select the physical optics solver for the reflector and the method of moment solver for the feed horn. Alternatively, MLFMM can be used to solve the full system directly. In summary, the four different options for analyzing the reflector are as follows:

Method 1)   Radiation pattern point source, PO solver for reflector
Method 2)   Radiation pattern point source, MLFMM solver for reflector
Method 3)   Full system: MoM solver for the horn, PO solver for reflector
Method 4)   Full system: MLFMM solver

How to select the PO solver for the analysis in FEKO was discussed earlier. For the radiation pattern point source excitation method, the far-field pattern of the source (a horn antenna in this design) is imported into FEKO, as shown in Figure 10-11a. The MLFMM solver can be selected in the solver setting (Figure 10-11b).

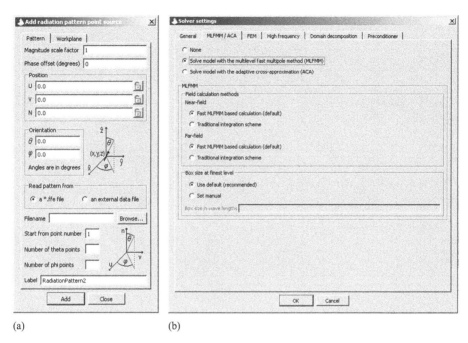

(a) (b)

FIGURE 10-11 ▪ Simulation setups in FEKO: (a) Point source radiation pattern. (b) Solver setting for MLFMM solution.

In the first three methods, the blockage effect caused by the feed horn will not be taken into account. For axisymmetric reflectors, feed blockage typically results in an increase in the first sidelobe and reduction of antenna gain. As such, if accurate analysis of the system is required, the latter option should be used. The radiation performance of this axisymmetric reflector is presented in Figure 10-12 using all four analysis methods. Despite some small differences, the general pattern shape obtained by the first three methods is in close agreement. On the other hand, the radiation pattern obtained by the fourth method is considerably different because blockage is not taken into account in the first three methods.

The radiation patterns obtained using these methods show that for axisymmetric reflectors it is imperative to properly account for the effects of blockage, and in general the fourth method is the most accurate.

The cross-polarized radiation pattern of this parabolic reflector, obtained by Method 4, is given in Figure 10-13. The maximum cross-polarization is observed in the diagonal planes.

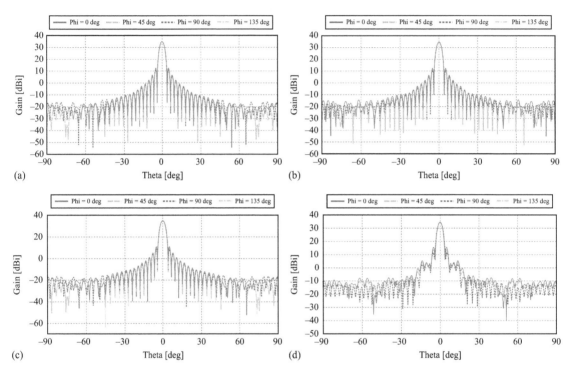

(a)   (b)   (c)   (d)

FIGURE 10-12 ▪ Co-polarized radiation patterns of the axisymmetric parabolic reflector: (a) Method 1.
(b) Method 2. (c) Method 3. (d) Method 4.

A summary of the performance of this parabolic reflector is also given in
Figure 10-14. The gain of the reflector is approximately 35 dB, and maximum
sidelobe level is about −19 dB down. As discussed earlier, the high sidelobe
level is attributed to the blockage effects of the feed horn antenna. The 3D

FIGURE 10-13 ▪
Cross-polarized
radiation patterns of
an axisymmetric
parabolic reflector
obtained using
Method 4.

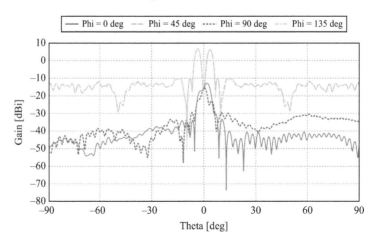

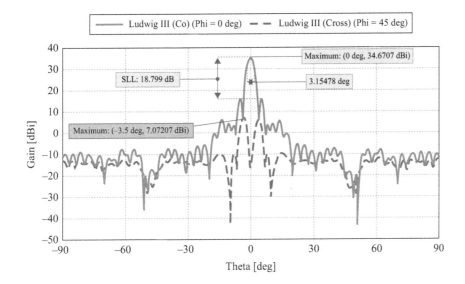

FIGURE 10-14 ▪
Radiation pattern of
an axisymmetric
parabolic reflector
antenna.

radiation pattern of this parabolic reflector computed using Method 4 is illustrated in Figure 10-15.

10.3.3 Offset Parabolic Reflectors

From the study presented in the previous section it is clear that one of the biggest problems associated with axisymmetric reflectors is the feed antenna's blockage effect. In general, for smaller reflectors the blockage effect is more significant but can be eliminated or significantly reduced using an offset reflector system. The offset system is more complicated; readers can see [7,8] for more details.

In this section we will study the performance of an offset parabolic reflector with the same aperture size as the axisymmetric design studied earlier. The cross sectional view of this parabolic reflector is given in Figure 10-16.

The reflector has the same aperture size (D), that is, 20λ, and is fed with the same corrugated conical horn antenna from section 9.6. The offset height (H) is set to be 11.5λ, as shown in Figure 10-16. The F/D in this offset design is set to 0.8 based on efficiency and edge taper consideration. The geometrical model of the reflector antenna in FEKO is also given in Figure 10-17.

All four methods can also be applied to analyzing this offset configuration; however, for brevity we will study only Methods 3 and 4. The radiation performance of this offset reflector is presented in Figure 10-18 using both methods of analysis. The main beam is correctly pointed in the $\theta = 0$ direction. Moreover, comparison between these patterns shows a very close agreement for

FIGURE 10-15 ■
3D radiation pattern
of an axisymmetric
parabolic reflector
antenna.

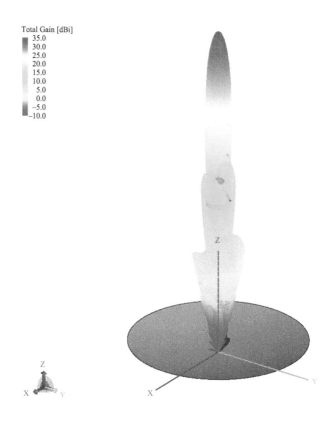

FIGURE 10-16 ■
Cross sectional view
of an X-band offset
parabolic reflector
antenna.

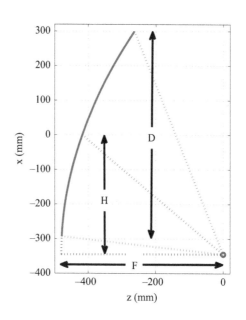

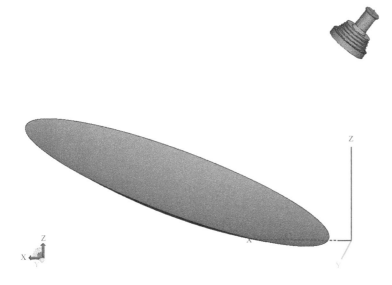

FIGURE 10-17 ▪
An offset parabolic
reflector antenna
modeled in FEKO.

both analysis methods. While in general Method 4 is still considered to be the most accurate solution, the PO solver has the advantage of significantly reducing computational time and resources, which makes it very effective for quickly analyzing offset reflector antennas.

A summary of the performance of this parabolic reflector is given in Figure 10-19, which shows that the gain of the reflector is approximately 35 dBi and maximum sidelobe level is about -22.5 dB down. Compared with the axisymmetric design, the gain of the offset configuration is increased by about 0.2 dB and the sidelobe level is decreased by about 3.7 dB. The 3D radiation pattern of this parabolic reflector is also presented in Figure 10-20.

▌ 10.4 | DESIGN AND ANALYSIS OF SPHERICAL REFLECTOR ANTENNAS

10.4.1 Basic Principles

In many high-gain applications it is desirable to scan the beam of the antenna. With parabolic reflectors, beam-scanning can be achieved by laterally displacing the feed antenna [24]; however, the performance is quite poor. The parabolic axisymmetric reflector is no longer a symmetric configuration when the feed is displaced from the focal point. On the other hand, the spherical reflector antenna, based on the concept of concave spherical mirrors in optics, has been

FIGURE 10-18 ▪
Radiation patterns of
an offset parabolic
reflector: (a) co-
polarized pattern
(Method 3).
(b) Cross-polarized
pattern (Method 3).
(c) Co-polarized
pattern (Method 4).
(d) Cross-polarized
pattern (Method 4).

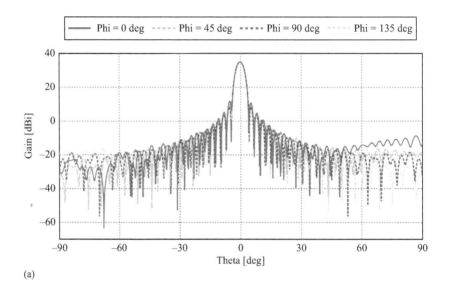

(a)

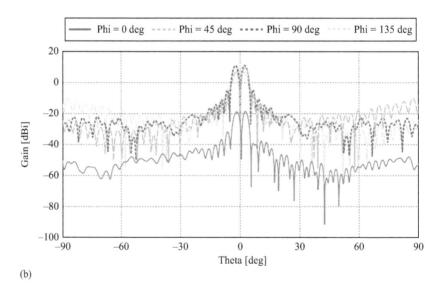

(b)

recognized for years as a suitable design for wide-angle high-gain beam-scanning
applications [25,26]. Because of its perfectly symmetric geometrical configuration, the spherical reflector can make an ideal wide-angle scanner, that is,
without radiation performance degradation. However, it is plagued by poor
inherent collimating properties due to spherical aberrations. Nevertheless,
different approaches have been introduced over the years to minimize these

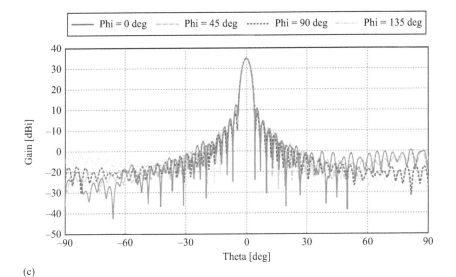

(c)

FIGURE 10-18 ▪
(*continued*).

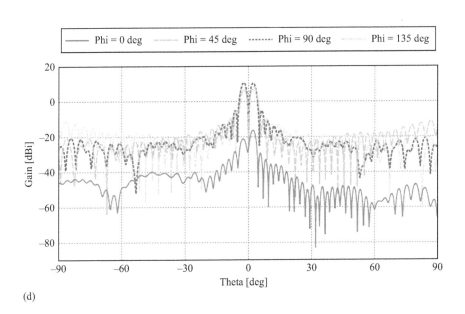

(d)

effects [25,26]. One of the simplest techniques is to use a restricted aperture and a reflector with a sufficiently large radius. The basic principle behind this approach is that for small angles the location of the focal point is independent of the angle of incidence, which means that all parallel rays that strike the spherical surface pass through the focal point (*F*). As such, in the restricted aperture design approach, one would illuminate only a small portion of the

FIGURE 10-19 ▪
Radiation pattern of
an offset parabolic
reflector antenna.

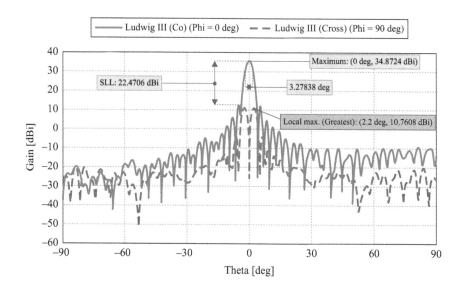

FIGURE 10-20 ▪
3D radiation pattern
of an offset parabolic
reflector antenna.

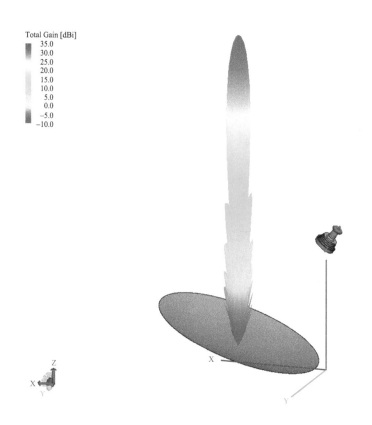

aperture (D_{ill}). The total usable physical aperture size (D) will then depend on the required scan range.

10.4.2 A Ka-band Spherical Reflector Antenna

In this section we will study the performance of a Ka-band spherical reflector antenna designed for 30° elevation coverage. To achieve a gain greater than 30 dBi, the diameter of illuminated aperture is selected to be 15λ. This would correspond to a sphere radius of 23.125λ. The geometrical model of the spherical reflector antenna in FEKO is given in Figure 10-21.

The next task is to determine the position of the feed antenna. The location of the focal optima in spherical reflectors (i.e., the paraxial focus) is roughly half the radius of the sphere. For a uniform taper, analytical expressions are available to accurately compute the position of focal optima [25], which for this configuration is 11.25λ. One important consideration in spherical reflectors is the blockage caused by the feed antenna. As discussed earlier, all reflected rays will pass through the focal point and the feed. For this small configuration, blockage would be quite large, so a point source model is used for the feed antenna. The radiation pattern model is a $\cos^q(\theta)$ function with $q = 4.6873$ to ensure that the taper at the virtual edge of each illuminated zone is -10 dB.

As discussed earlier, to achieve beam-scanning we move the feed along a circular arc to illuminate the appropriate portions of the reflector surface. The center of this arc is the same as the sphere center. The currents on the surface of the spherical reflector for a few different cases are shown in Figure 10-22 and clearly illustrate the illumination requirement for beam-scanning with spherical reflectors. Depending on which portion of the reflector

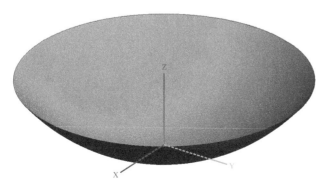

FIGURE 10-21 ▨
A spherical reflector antenna modeled in FEKO.

FIGURE 10-22 ▪
Currents on the
surface of a
spherical reflector
obtained using the
PO solver in FEKO:
(a) $\theta = 0°$. (b) $\theta = 15°$.
(c) $\theta = 30°$.

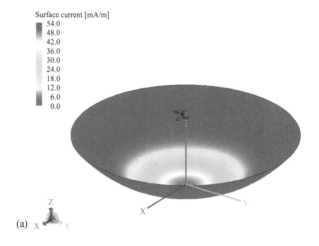

(a)

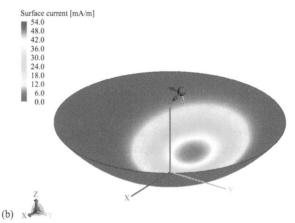

(b)

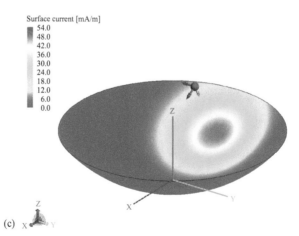

(c)

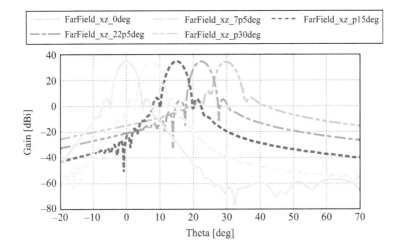

FIGURE 10-23 ▪
Beam-scanning
performance of a
spherical reflector
antenna.

is active, a beam will be generated that is normal to that local illuminated surface.

The perfect symmetrical configuration of the spherical reflector means that a similar illumination can be realized on the aperture for every azimuth direction. As such the spherical reflector can provide a full azimuth scan. Remember that the amount of elevation coverage depends on the design, and very wide angle coverage is possible. The system studied here was designed for 30° elevation coverage, and the simulated beam-scanning performance is shown in Figure 10-23. When the feed is placed as shown in Figure 10-22a, the beam is in the broadside direction (i.e., the blue curve in Figure 10-23). As the feed moves along the displacement path, the beam is scanned. When the feed is placed as shown in Figure 10-22c, the beam is scanned 30° from broadside direction (i.e., the green curve in Figure 10-23).

Very good beam-scanning performance is obtained with this design. The gain of the reflector is almost constant for all scan angles. The side-lobe level slightly increases as we approach the maximum scan angles, but it is still more than 20 dB down for all cases.

EXERCISES

(1) *Designing a Corner Reflector Antenna.* Using the same dimensions as given for the 90° corner reflector in section 10.2, design a corner reflector antenna with a 60° subtended angle. Study the effect of feed to apex distance for this configuration. What is the optimum position for the feed antenna?

(2) *Beam-Scanning with a Corner Reflector Antenna.* Using the same dimensions as given for the 90° corner reflector in section 10.2, study the

effect of lateral feed displacement. Move the center of the dipole antenna along the direction of the reflector length, i.e. z-axis, while tilting the dipole about the y-axis such that the main beam still points to the center of the reflector. Compare the radiation pattern of this scanned beam with the nonscanned case for scan angles of $10°$ and $20°$.

(3) ***Designing an Axisymmetric Parabolic Reflector Antenna.*** Design a parabolic reflector antenna with a 30 dBi gain for Ku-band satellite. The center frequency is 12.45 GHz. For the feed antenna you can use a pyramidal horn antenna as described in Chapter 9. Select the F/D such that the maximum edge taper is below -10 dB.

(4) ***Designing an Offset Parabolic Reflector Antenna.*** Repeat exercise 3 for an offset configuration. Chose appropriate values for the offset system such that the feed blockage is completely eliminated.

Antenna Arrays

Chapter Outline

11.1 | INTRODUCTION

In the previous chapters, several types of antenna configurations were reviewed and analyzed using FEKO electromagnetic software. With the exception of reflectors, all of these antennas have a relatively wide radiation pattern and consequently a low to moderate gain. For long-distance communications, we need to increase the gain of these antennas. To this end, we can employ multiple antennas that form an array (called array elements), each of which radiates a beam, and with a proper array setup the gain of the antenna array can be increased. Moreover, by controlling the phase of the array elements, the array can also scan the main beam of the antenna array.

An antenna array offers several degrees of freedom that can be used to shape the overall pattern of the antenna. The first task in designing an array is selecting the geometrical configuration, the most common types of which are linear and planar (circular or rectangular). Nonplanar arrays such as cylindrical and spherical are also desirable in certain applications. Once the geometry of the array is chosen, the next task is to select the relative placement of the elements in the array.

In most antenna array configurations, the elements are placed with an equal spacing. Once the array architecture is designed, the third task is to determine the excitations (magnitude and phase) for the array elements. These three factors are used to shape the overall pattern of the array. Another important factor in designing the antenna array is selecting the type of antenna that will be used for the elements of the array, which in most cases are identical. Although almost any type of antenna can be used, the common types of elements used for array antennas are dipoles and microstrip patches. In this chapter we will study several types of antenna array configurations using dipole and microstrip patch elements.

11.2 | BASICS OF ANTENNA ARRAYS

In this section we will briefly discuss the governing dynamics of antenna arrays and the fundamental equations used for radiation analysis. For simplicity we will focus only on linear arrays with identical elements and uniform element spacing, but interested readers are encouraged to study more advanced array configurations [27–30].

The radiation pattern of an antenna array depends on two almost independent parameters: (1) the radiation pattern of a single element in an unbounded medium known as the element pattern; and (2) the radiation pattern of the array when isotropic point sources are placed instead of the elements. The latter is known as the array factor. The radiation pattern of the total array can then be computed using pattern multiplication for arrays of identical elements, which is given by [7]

$$E \ (total) = E \ (single \ element) \times Array \ Factor. \tag{11-1}$$

For the element pattern, it is desirable to select an element that provides a suitable pattern for the array. For example, if an array is to have the main beam pointing in $\theta = 0°$ direction, a dipole antenna placed along the z axis will not be suitable since it has a null in that direction. On the other hand, a horizontal dipole (placed in the x-y plane) or a patch antenna placed in the x-y plane is a suitable choice for such a design. The radiation patterns of several different antennas were studied in previous chapters, and in general all these antennas can be used as the array elements.

Now let's focus on the second term in equation (11-1): the array factor. For an N-element linear array of identical elements and equal element spacing (d), oriented along the z axis, and each with a progressive phase (β), the array factor can be written as

$$AF = a_1 + a_2 e^{+j(kd \cos \theta + \beta)} + a_3 e^{+2j(kd \cos \theta + \beta)} + \cdots + a_N e^{+j(N-1)(kd \cos \theta + \beta)}$$

$$= \sum_{n=1}^{N} a_n e^{+j(n-1)(kd \cos \theta + \beta)}, \tag{11-2}$$

where θ is the far-field observation angle, a is the excitation of each element, and the phase of the array elements increases by β radians from one element to the next. k is the wavenumber which is defined as $2\pi/\lambda$. If the excitation is uniform (equal amplitude), the array factor can be simplified into

$$AF = \sum_{n=1}^{N} e^{+j(n-1)(kd\cos\theta+\beta)}. \qquad (11\text{-}3)$$

It can be shown [7] that if the reference point is set to be the physical center of the array, (11-3) can be simplified into

$$AF = \frac{\sin(N\psi/2)}{\sin(\psi/2)}, \quad \psi = kd\cos\theta + \beta. \qquad (11\text{-}4)$$

This equation permits us to determine the maximums of the array factor in terms of ψ. The first maximum of the array factor in (11-4) occurs when

$$\psi = kd\cos\theta + \beta = 0. \qquad (11\text{-}5)$$

The progressive phase of the array elements is then used to scan the main beam of the antenna. Therefore, if the required beam direction ($\theta = \theta_m$) is known, one can solve (11-5) and determine the progressive phase. For example, to have a beam normal to the axis of the array ($\theta = 90°$), that is, broadside direction, we have

$$\psi = kd\cos\theta + \beta|_{\theta=90°} = \beta = 0. \qquad (11\text{-}6)$$

On the other hand, to scan the main beam of the array to $\theta = 0°$ (known as end-fire), we have

$$\psi = kd\cos\theta + \beta|_{\theta=0°} = kd + \beta = 0. \qquad (11\text{-}7)$$

Therefore, the progressive phase will be given by

$$\beta = -kd. \qquad (11\text{-}8)$$

For the general case where $\theta = \theta_m$, the progressive phase is given by

$$\beta = -kd\cos\theta_m. \qquad (11\text{-}9)$$

One objective in most cases is to avoid having multiple maxima (in addition to the main beam), known as grating lobes. This condition is dependent on both the element spacing and the scan angle and is calculated using equations (11-4) and (11-9). To suppress all grating lobes for a linear array, the maximum element spacing is given by

$$d_{\max} = \frac{\lambda}{1 + |\cos\theta_m|}. \qquad (11\text{-}10)$$

The materials covered in this section are the basic concepts with which antenna engineers must be familiar to design an efficient array architecture. In the next section we will study several different configurations of arrays using dipole antennas for the array elements.

11.3 | TWO-ELEMENT DIPOLE ARRAYS

Now let us start our studies on array antennas with a simple example of a two-element dipole array. We use the ultra high frequency (UHF) dipole antenna designed in Chapter 2, which has a total length of 0.4823λ for the array element. The distance from the center of one dipole to the other (element spacing) is set to 0.5λ, and the centers of the two dipoles are placed along the z axis. Two different configurations are studied for the dipoles (Figure 11-1): both dipoles oriented in the z direction (Figure 11-1a); and both oriented along the x axis (Figure 11-1b). In the first case the dipoles are not in physical contact.

FIGURE 11-1 ■
Geometrical model of a two-element dipole array, placed along the z axis.

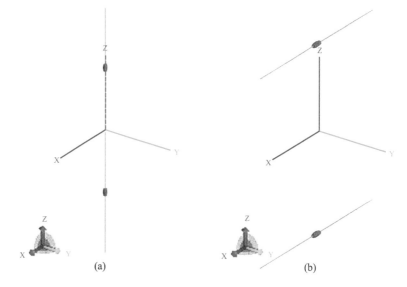

(a)   (b)

To analyze this array's performance, we first determine the array factor using equation (11-2). Both arrays have an equal phase and amplitude (i.e. $a_1 = a_2, \beta = 0$). The array factor for both configurations can then be written as

$$AF = \frac{\sin(\pi \cos \theta)}{\sin\left(\dfrac{\pi}{2} \cos \theta\right)}. \tag{11-11}$$

A normalized plot of this array factor is given in Figure 11-2. The array factor has a maximum in the $\theta = 90°$ direction and a null (zero) in $\theta = 0°$ and $180°$ directions.

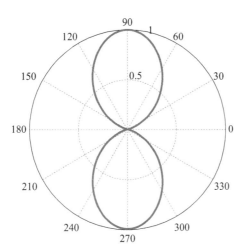

FIGURE 11-2 ▪
Array factor of a
two-element array
with uniform
excitation and
$d = \lambda/2$.

Now let's look at the element patterns. For case (a), the radiation pattern of a dipole placed along the z axis has maximums at $\theta = 90°$ and $270°$ directions and nulls (zeros) at $\theta = 0°$ and $180°$ directions (Chapter 2, Figure 2-6). As discussed earlier, for arrays with identical elements the radiation pattern of the array is a direct multiplication of the element pattern by the array factor (11-1). In this case since the peaks (and minimums) of the array factor and element pattern are in the same directions, the radiation pattern of this dipole array is expected to be similar to the dipole. The simulated radiation patterns are given in Figure 11-3.

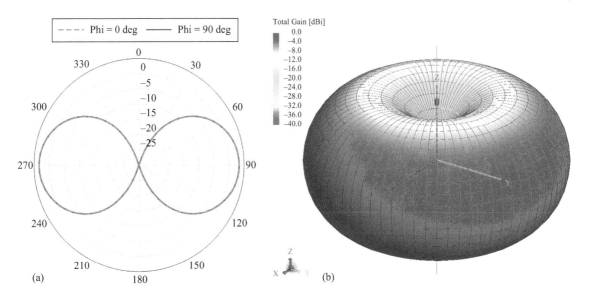

FIGURE 11-3 ▪ Radiation patterns for the two-element dipole array in Figure 11-1a: (a) 2D. (b) 3D.

For case (b), however, the radiation patterns would be quite different. The dipoles are placed along the x axis, so the maximum element patterns would be observed at $\theta = 0°$ and $180°$ directions, and the nulls would be at $\theta = 90°$ and $270°$. As such this array would observe multiple nulls associated with both the element pattern and the array factor. In the $\varphi = 0°$ plane (x-z plane), the nulls due to the array factor will be observed at $\theta = 0°$ and $180°$ and the nulls due to the element pattern will be observed at $\theta = 90°$ and $270°$. In the $\varphi = 90°$ plane (y-z plane), the nulls due to the array factor will still be observed at $\theta = 0°$ and $180°$, but the nulls due to the element pattern do not occur in this plane. In fact, the peak of the array pattern is observed in this plane along the $\theta = 90°$ and $270°$ directions. The simulated radiation patterns for this case are given in Figure 11-4.

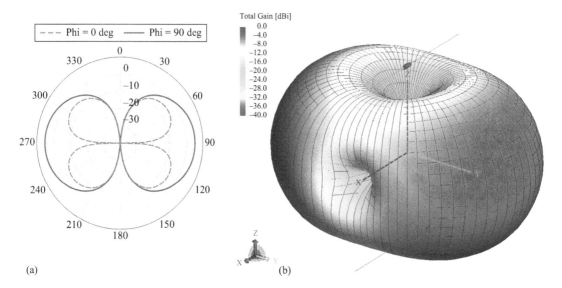

(a)   (b)

FIGURE 11-4 ▪ Radiation patterns for the two-element dipole array in Figure 11-1b: (a) 2D. (b) 3D.

This simple example of a two-element dipole array clearly shows the importance of properly understanding an antenna array's radiation mechanism for an efficient design. The total radiation pattern of an array is directly influenced by both the element pattern and the array factor, and as such proper orientation of the elements of an array is of prominent importance in achieving the desired radiation pattern.

It is important to emphasize here that in both cases—and in general for any array configuration—the proximity of the elements of the array causes some coupling between them. This phenomenon is known as mutual coupling and has to be taken into account when designing the antenna array. Here we are referring to the approximation made in equation (11-1), where the element pattern is obtained in unbounded space, and it is not the same as the pattern of a dipole when it is in an array. Nonetheless, this approximation is quite valid as demonstrated with the aforementioned example. In the next section we will study the radiation patterns of linear arrays with multiple elements.

11.4 | *N*-ELEMENT UNIFORM-AMPLITUDE LINEAR DIPOLE ARRAYS

11.4.1 Element Spacing

As discussed in section 11.3, the array factor of a linear array antenna with uniform excitation can be computed using equations (11-3) or (11-4). Without the loss of generality we will investigate the radiation performance of five-element dipole array antennas by studying the effect of element spacing on their radiation performance. All arrays are designed for a broadside beam (i.e., $\beta = 0$). We consider three different element spacings for this array: $d = \lambda/4$, $\lambda/2$, and λ. Their array factors are given in Figure 11-5.

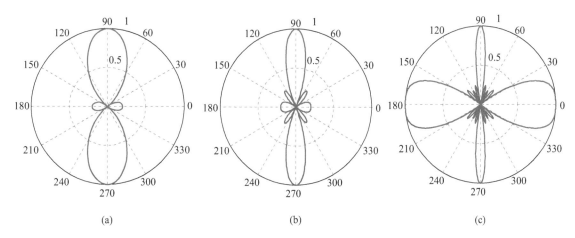

FIGURE 11-5 ▪ Array factors for five-element linear arrays: (a) $d = \lambda/4$. (b) $d = \lambda/2$. (c) $d = \lambda$.

When the element spacing is smaller than λ, the array factor has its maximums at $\theta = 90°$ and $270°$ directions, that is, the broadside direction. However, when the element spacing is equal to λ, grating lobes appear at $\theta = 0°$ and $180°$. The directions of the grating lobes can also be computed directly using equation (11-10), that is,

$$|\cos \theta_m| = \pm 1 \Rightarrow \theta_m = 0°, 180°. \qquad (11\text{-}12)$$

As discussed before, in general this condition is undesirable for an array, but the aim of this study is to observe these effects on the radiation pattern. In addition, as the element spacing increases the beamwidth of the array factor decreases because the electrical size of the antenna is increasing.

For this dipole array, we place the center of each dipole element along the z axis, oriented parallel to the x axis as shown in Figure 11-6. As we saw in section 11.3, with such a configuration the main beam of this array will be observed in the $\varphi = 90°$ plane (y-z plane), $\theta = 90°$ and $270°$ directions. The radiation patterns of these dipole array antennas are given in Figure 11-7.

Comparing the radiation patterns for these three antenna arrays clearly shows that the spacing of the elements of the array changes the radiation pattern. In fact, when the element spacing changes from $\lambda/4$ to $\lambda/2$, the beamwidth of the antenna decreases and the gain increases from 3.74 to 9.07 dBi. However, when the element spacing changes from $\lambda/2$ to λ, grating lobes appear in the array radiation pattern. At $\theta = 0°$ and $180°$, they take the energy away from the main beams of the array (at $\theta = 90°$ and $270°$). Consequently, while the aperture size has doubled, the peak gain has decreased. The maximum gain for this array is 8.27 dB.

This study reveals the importance of element spacing in the performance of the array. As we observed here, grating lobes will appear in the array pattern if

FIGURE 11-6 ▪
Geometrical setup of five-element dipole array with $d = \lambda/2$.

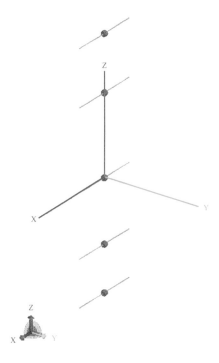

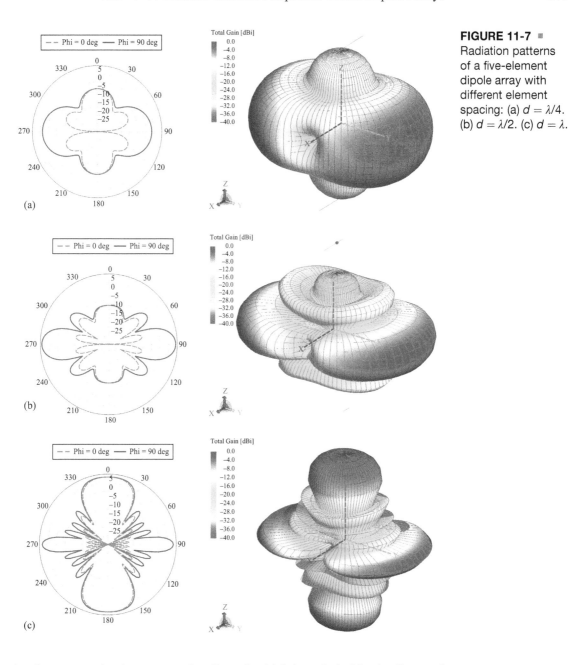

FIGURE 11-7 ▪
Radiation patterns of a five-element dipole array with different element spacing: (a) $d = \lambda/4$. (b) $d = \lambda/2$. (c) $d = \lambda$.

the element spacing is not properly adjusted, which is undesirable. As discussed earlier, a generation of grating lobes also depends on the beam direction of the array. In the next section we will study the performance of linear arrays with scanned beams.

11.4.2 Beam Scanning

To scan the main beam of the array, we need to add a progressive phase to the elements of the array antenna as given by equation (11-9). Here we will study the scanning performance of the same five-element dipole array in the previous section. To avoid grating lobes in the array radiation pattern, we have to satisfy the condition of equation (11-10). In general, the element spacing can be selected based on the scan angle. However, to ensure that no grating lobe is observed for all scan angles ($\theta_m =$ from 0° to 90°), it follows from (11-10) that the maximum element spacing should not exceed $\lambda/2$. Therefore, we select $d = \lambda/2$ for our five-element dipole array and study its scan performance, with the provision that no grating lobes will appear as we scan the beam.

For an element spacing of $\lambda/2$, the progressive phase required to scan the beam is computed using (11-9) as

$$\beta = -\pi \cos \theta_m. \tag{11-13}$$

To set up this simulation in FEKO we assign the appropriate progressive phase to each element of the array. The excitation setup for the second element of the array is shown in Figure 11-8.

FIGURE 11-8 ▪ The excitation setup for a linear scanning dipole array in FEKO.

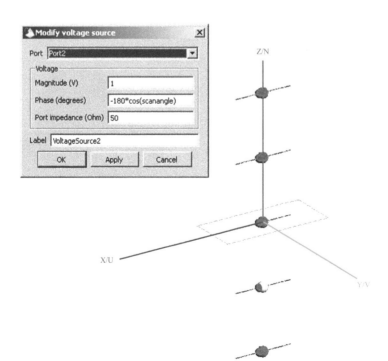

As we saw in the previous cases, for this configuration when the progressive phase is zero the main beam will be at $\theta_m = 90°$ and 270°. Let's consider the beam at 90° and add progressive phase shifts such that the main beam shall be scanned from 90° to 0° with a 30° step. The required element phases for these four cases are as follows:

No scan $(\theta_m = 90°)$: $0°, 0°, 0°, 0°, 0°$
30° scan $(\theta_m = 60°)$: $0°, -90°, -180°, -270°, -360°$
60° scan $(\theta_m = 30°)$: $0°, -155.88°, -311.77°, -467.65°, -623.54°$
90° scan $(\theta_m = 0°)$: $0°, -180°, -360°, -540°, -720°$

Once the phases of the array elements are known, the associated array factor can be computed using equation (11-3). The array factors for these four scan cases are shown in Figure 11-9.

By properly exciting the elements, the array factor is correctly scanned to the required direction, but for wider scan angles (e.g., Figure 11-9c) the radiation pattern is somewhat distorted; in practice linear (or planar) arrays are typically used only for scan angles less than 60°.

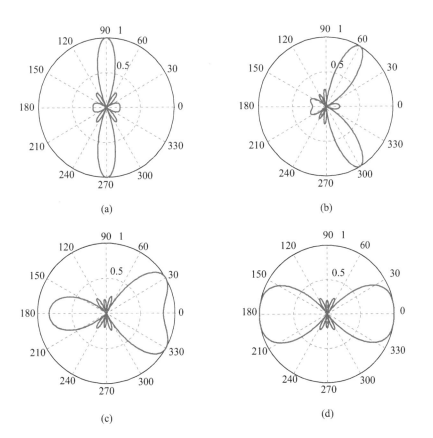

(a)

(b)

(c)

(d)

FIGURE 11-9 ▪
Array factors for
a five-element
linear array with
different progressive
phases: (a) $\beta = 0°$.
(b) $\beta = -90°$.
(c) $\beta = -155.88°$.
(d) $\beta = -180°$.

FIGURE 11-10 ◾
Radiation patterns
for a five-element
linear array of
dipoles with different
progressive phases:
(a) $\beta = 0°$.
(b) $\beta = -90°$.
(c) $\beta = -155.88°$.
(d) $\beta = -180°$.

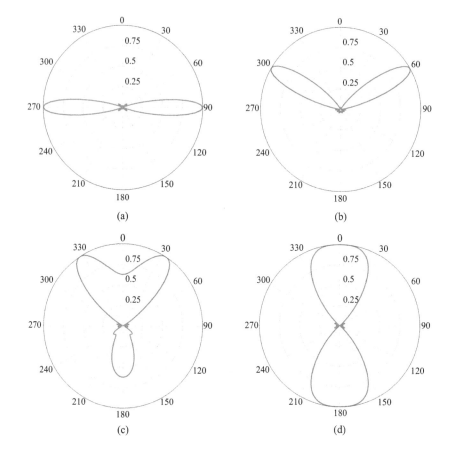

Now let's see the performance of the scanning linear dipole array antenna of Figure 11-8. The radiation patterns in the y-z plane are given in Figure 11-10.

The radiation patterns of the antenna array are close to what is expected; in other words, the beam is almost scanned to the required direction. However, for example in Figure 11-10c, the main beam is actually pointing at 35° rather than at the required 30° because of the effect of the element pattern on the performance of the array. Nonetheless, by properly controlling the element phase shift, this antenna array demonstrated a complete scan coverage of 90°.

▌ 11.5 ▐ N-ELEMENT LINEAR DIPOLE ARRAYS WITH NONUNIFORM AMPLITUDE

Several examples of linear arrays were presented in the previous sections, but all of their elements were excited with the same amplitude. In general, uniform arrays achieve the largest directivity and therefore the narrowest beamwidth.

While this is certainly an advantage, in many applications we prefer to reduce the antenna's sidelobe level. Nonuniform element amplitude in array configurations provides the means to achieve this, although with some compromise in antenna directivity. In other words, if we aim at reducing the sidelobe level, the directivity of the antenna will also be reduced.

Various amplitude distribution designs have been studied for array antennas, such as binomial, Dolph-Tschebyscheff, and Taylor. In all of them, the element amplitude is large at the center of the array and decays toward the edge elements. Here we will study only the radiation performance of the five-element dipole array of section 11.4, with binomial and Dolph-Tschebyscheff amplitude distributions.

Binomial designs typically have the smallest sidelobes, and for an element spacing of $\lambda/2$ they have none. For the five-element array, the amplitude distribution for a binomial design is

Binomial: 0.1667, 0.6667, 1.0, 0.6667, 0.1667.

The Dolph-Tschebyscheff distribution is basically a compromise between uniform and binomial distributions. Its great advantage is that the element amplitudes can be determined based on the required sidelobe level. For the five-element array, the amplitude distribution for a Dolph-Tschebyscheff design with a sidelobe level of -20 dB is

Dolph-Tschebyscheff: 0.5176, 0.8326, 1.0, 0.8326, 0.5176.

The normalized radiation patterns in the yz plane for both of these cases are given in Figure 11-11. These results are to be compared with Figure 11-7b (i.e., the uniform design). As expected, the binomial design completely suppresses the sidelobes; however, this results in a much broader beam. On the other hand, the Dolph-Tschebyscheff design is achieving a narrower beam width at the expense of some compromise in sidelobe level.

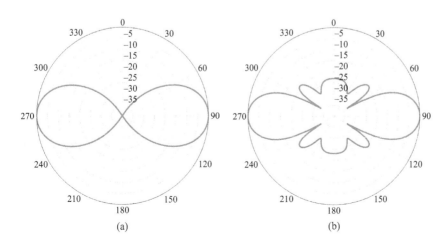

(a)   (b)

FIGURE 11-11 ◾ Radiation patterns for a five-element linear array with different amplitude distributions: (a) Binomial. (b) Dolph-Tschebyscheff.

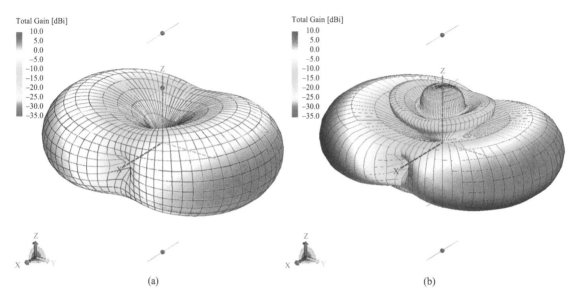

(a)   (b)

FIGURE 11-12 ■ 3D radiation patterns for a five-element linear array with different amplitude distributions: (a) Binomial. (b) Dolph-Tschebyscheff.

The 3D radiation patterns for both these designs are provided in Figure 11-12. It is important to emphasize that in most cases array antennas are designed with some form of amplitude taper to control the sidelobe level. The results presented in this section are the ideal amplitude distributions for the corresponding design. In practice, though, power dividers must be designed for the array elements, which typically limit the control level on the element amplitude.

▮ 11.6 ▏ PLANAR ANTENNA ARRAYS

A planar array configuration offers additional variables that can be used to control the pattern of the antenna array. Typically the elements are placed along a rectangular grid, although other configurations such as polar are also possible. Compared with linear arrays, planar configurations can provide more symmetrical patterns (unlike the fan beams of linear arrays), and they can also scan the main beam of the array to any point in space. In this section we will study a few different configurations of planar array antennas.

11.6.1 A 4 × 4 Dipole Array

The same UHF dipole studied in the previous sections is used here to design a planar array configuration. A sixteen-element dipole array (4 × 4) is designed and placed on the x-y plane as shown in Figure 11-13. All elements are excited with the same amplitude and phase. As we saw in sections 11.3 and 11.4, the array factor of such an array will be normal to its plane. Therefore, with the config-uration of Figure 11-13 the main beam is expected to point at $\theta = 0°$ and $180°$.

The radiation patterns of this planar dipole array including mutual coupling effects are provided in Figure 11-14. As expected, the main beams are correctly pointing in the directions normal to the aperture of the array. Moreover, the radiation pattern pointing at $\varphi = 0°$ and $90°$ planes are almost similar.

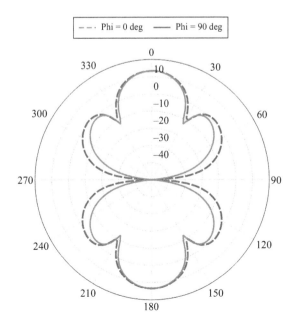

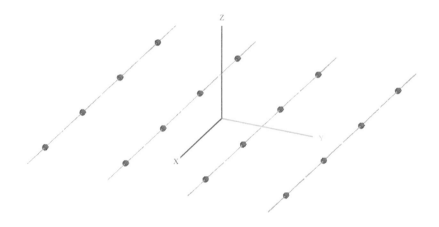

FIGURE 11-13 ■ A sixteen-element planar dipole array antenna.

FIGURE 11-14 ■ 2D radiation patterns of a sixteen-element planar dipole array antenna.

FIGURE 11-15 ■ 3D radiation pattern of the sixteen-element planar dipole array antenna.

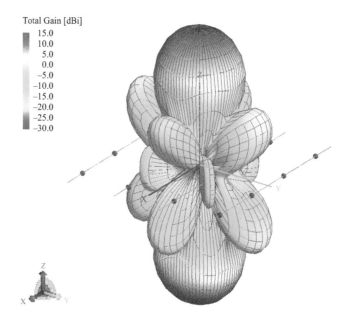

As discussed earlier, this almost symmetric radiation pattern is a very interesting feature of planar array configurations. The 3D pattern of this dipole array antenna is also shown in Figure 11-15. The gain of this dipole array is 13.35 dBi at 300 MHz.

One disadvantage of these dipole arrays is that, since the dipole has a pattern that is azimuthally symmetric, two main beams will be generated with such an array. One way to suppress the second beam (e.g., the beam at $\theta = 180°$) is to place a ground plane beneath the dipole array. Another approach is to use an element other than a dipole that has no radiation in the lower hemisphere. An ideal example is the microstrip patch, which will be studied in the next section.

11.6.2 A 2×2 Microstrip Patch Array

In this section, we use a microstrip patch antenna as the array element. The array elements are to be fed with a microstrip transmission line, so the inset-fed microstrip patch configuration (Chapter 4) and design of a microstrip array antenna can be used to achieve a gain in the order of 10 dBi. Typically for such gain a 2×2 microstrip patch array is sufficient. The array is designed for HiperLAN operation (i.e., 5.8 GHz).

A 1.27 mm thick Rogers 6006 laminate (dielectric constant = 2.2, loss tangent = 0.0019) is selected for the design, and the element spacing is set to 25 mm. The patch size is 10.1×13.9 mm^2, and the inset is 3.4 mm. The size of the ground plane is 50.8×50.8 mm^2. The inset feed line provides an impedance of 50 Ω, and a feed network is designed such that the entire antenna can be fed with one 50 Ω port. The geometrical model of this microstrip patch array is shown in Figure 11-16.

The input impedance and $|S_{11}|$ at the pin feed port is illustrated in Figure 11-17. The antenna is well matched at the band of interest, and the center frequency

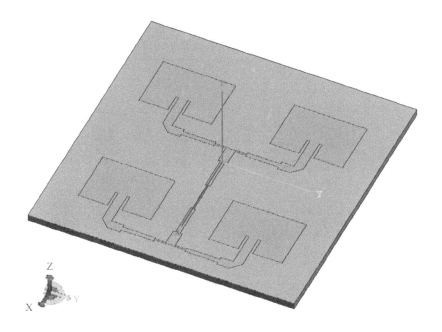

FIGURE 11-16 ▪
A planar 2 × 2 microstrip patch array antenna.

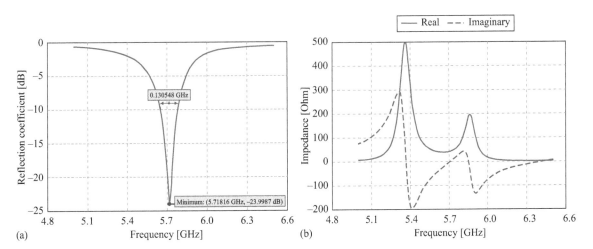

FIGURE 11-17 ▪ The performance of a planar 2 × 2 microstrip patch array antenna as a function of frequency: (a) Input impedance. (b) $|S_{11}|$.

is 5.72 GHz. The current distribution on the array aperture is given in Figure 11-18. As expected, all patch elements have an almost similar current distribution.

The radiation patterns of this planar microstrip antenna array at 5.72 GHz are shown in Figure 11-19 and Figure 11-20. As expected, the radiation pattern

FIGURE 11-18 ■
Current distribution on a planar 2 × 2 microstrip patch array antenna at 5.72 GHz.

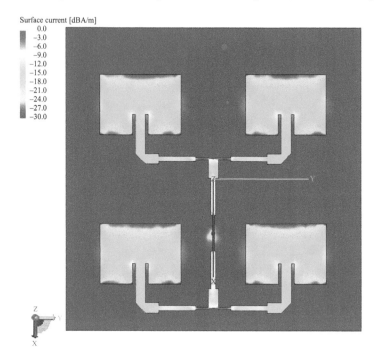

FIGURE 11-19 ■
2D radiation patterns of a planar 2 × 2 microstrip patch array antenna at 5.72 GHz.

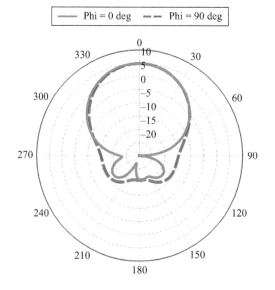

Total Gain [dBi]

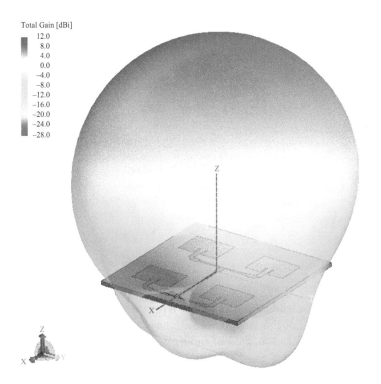

FIGURE 11-20 ■
3D radiation pattern
of a planar 2 × 2
microstrip patch
array antenna at
5.72 GHz.

is almost completely in the top hemisphere, which is a significant improvement over the dipole array configuration studied earlier. Moreover, a very good symmetric beam is obtained with this array (Figure 11-19). The peak gain of this antenna is 9.87 dBi.

11.6.3 Microstrip Patch Reflectarray

Reflectarray antennas imitate conventional parabolic reflectors but have a low profile, low mass, and flat surface. This relatively new hybrid design combines the numerous advantages of both printed phased arrays and parabolic reflectors [31] and has emerged as the new generation of high-gain antennas.

Here we consider a Ka-band reflectarray with a circular aperture and a diameter of 14.5λ at the design frequency of 32 GHz. The feed is positioned at $X_{feed} = -45.90$ mm, $Y_{feed} = 0$ mm, and $Z_{feed} = 98.44$ mm based on the coordinate system in Figure 11-21. The element phase is designed to generate a beam in the direction of $(\theta, \varphi) = (25°, 0°)$. For the reflectarray phasing elements, the variable size square patches are selected from the S-curve data in [31]. The 609-element reflectarray antenna is modeled using the commercial electromagnetic software FEKO [9]. For the excitation of the reflectarray, a point source feed model with a $\cos^{6.5}\theta$ radiation pattern is used. The advantage of using a point

FIGURE 11-21 ■
(a) Top view of a reflectarray. (b) 3D radiation pattern of the reflectarray at 32 GHz.

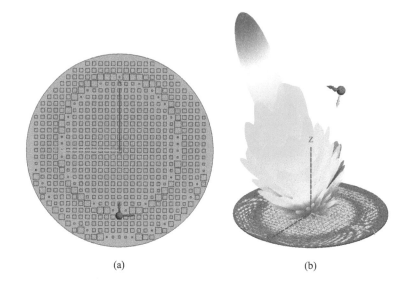

(a)   (b)

source rather than a feed horn here is that a point source model does not have a blockage aperture. For this design 568,435 unknown basis functions have to be calculated for the FEKO method of moments (MoM) solution. Considering the large number of unknowns, the multilevel fast multipole method (MLFMM) solver in FEKO was chosen for this simulation. The geometry of the reflectarray antenna modeled in FEKO and the simulated 3D radiation pattern are presented in Figure 11-21.

It is important to note here that the full-wave simulation required 29.56 GB of memory with a CPU time of 26.97 hours on an eight-core 2.66 GHz Intel(R) Xeon(R) E5430 computer.

EXERCISES

(1) *Uniform Linear Loop Arrays.* Using the same geometrical setup as in section 11.3, study the performance of two-element loop array antennas. Consider two configurations: (1) when the plane of the two loops is parallel and (2) when it is orthogonal.

(2) *Dolph-Tschebyscheff Linear Dipole Arrays.* Using the same geometrical setup as in section 11.5, design a ten-element linear dipole array with Dolph-Tschebyscheff amplitude distribution. Consider designs with three sidelobe levels: −15 dB, −20 dB, and −25 dB. Compare the beam width and gain of these array antennas and discuss your observations.

(3) *Linear Microstrip Array.* Using the same material and patch dimensions of section 11.6, design a four-element linear patch array, such that the main

beam points at $\theta = 0°$. Compare the performance of this linear array with the planar four-element design of section 11.6. Which array has a higher gain?

(4) ***Planar Dipole Arrays with Ground Plane.*** Study the performance of the planar dipole array of section 11.6, when an infinite ground plane is added beneath the array. Study the effect of the ground plane position on the performance of this array. At what position does this array achieve the maximum directivity?

References

[1] http://www.newscotland1398.net/nfld1901/marconi-nfld.html.

[2] A. H. Systems, Inc., http://www.ahsystems.com.

[3] D-link Corporation, http://www.dlink.com.

[4] ETS-Lindgren, http://www.ets-lindgren.com.

[5] http://en.wikipedia.org/wiki/File:Erdfunkstelle_Raisting_2.jpg.

[6] http://encyclopedia2.thefreedictionary.com/MIMO.

[7] C. A. Balanis, *Antenna Theory: Analysis and Design*, 3rd ed., John Wiley & Sons Inc., 2005.

[8] W. L. Stutzman and G. A. Thiele, *Antenna Theory and Design*, 3rd ed., John Wiley & Sons Inc., 2012.

[9] FEKO Comprehensive Electromagnetic Solutions, v6.2, EM Software & Systems Inc., 2013.

[10] K. S. Nikita, G. S. Stamatakos, N. K. Uzunoglu, and A. Karafotias, "Analysis of the interaction between a layered spherical human head model and a finite-length dipole," in *IEEE Trans. Microwave Theory and Tech.*, vol. 48, no. 11, pp. 2003–2013, 2000.

[11] A. Z. Elsherbeni, J. Colburn, Y. Rahmat-Samii, and C. D. Taylor, Jr., "On The Interaction of Electromagnetic Fields With a Human Head Model Using Computer Visualization", Oristaglio, M. and Spies, B., Ed., *Three-Dimensional Electromagnetics: Society of Exploration Geophysicists (SEG)*, pp. 671–684, 1999.

[12] H.-T. Hsu, J. Rautio, and S.-W. Chang, "Novel planar wideband omni-directional quasi log-periodic antenna," *Proceedings of the Asia Pacific Microwave Conference (APMC)*, Suzhou, China, 2005.

[13] K. F. Lee and K. M. Luk, *Microstrip Patch Antennas*, London, Imperial College Press, 2010.

[14] T. Huynh and K. F. Lee, "Single-layer single-patch wideband microstrip antenna," in *Electron. Lett.*, vol. 31, no. 16, pp. 1310–1312, 1995.

[15] K. F. Lee, K. M. Luk, K. F. Tong, Y. L. Yung, and T. Huynh, "Experimental study of the rectangular patch with a U-shaped slot," in *IEEE Antennas Propag. Soc. Int. Symp. Dig.*, vol. 1, pp. 10–13, 1996.

[16] K. F. Tong, K. M. Luk, K. F. Lee, and R. Q. Lee, "A broad-band U-slot rectangular patch antenna on a microwave substrate," in *IEEE Trans. Antennas Propag.*, vol. 48, no. 6, pp. 954–960, 2000.

[17] F. Yang, X. Zhang, X. Ye, and Y. Rahmat-Samii, "Wideband E-shaped patch antennas for wireless communications," in *IEEE Trans. Antennas Propag.*, vol. 49, no. 7, pp. 1094–1100, 2001.

[18] D. M. Pozar, *Microwave Engineering*, 3rd ed., John Wiley & Sons Inc., 2005.

[19] J. D. Kraus and R. J. Marhefka, *Antennas for All Applications*, 3rd ed., McGraw-Hill, 2001.

[20] P. D. Potter, "A new horn antenna with suppressed sidelobes and equal beam-widths," JPL Technical Report No. 32-354, 1963.

[21] A. D. Oliver, P. J. B. Clarricoats, A. A. Kish, and L. Shafai, *Microwave Horns and Feeds*, Institution of Electrical Engineers, IEE Electromagnetic Waves Series, 1994.

[22] J. L. Volakis, *Antenna Engineering Handbook*, 4th ed., McGraw-Hill, 2007.

[23] Antenna Magus v4.1, 2013.

[24] Y. Rahmat-Samii, "Reflector antennas", in Y.T. Lo and S.W. Lee (eds). *Antenna Handbook: Theory, Applications, and Design*, Van Nostrand Reinhold, 1988.

[25] T. Li, "A study of spherical reflectors as wide-angle scanning antennas," in *IEEE Trans. Antennas Propag.*, vol. AP-7, no. 3, pp. 223–226, 1959.

[26] R. Spencer and G. Hyde, "Studies of the focal region of a spherical reflector: Geometric optics," in *IEEE Trans. Antennas Propag.*, vol. AP-16, no. 3, pp. 317–324, 1968.

[27] R. S. Elliot, *Antenna Theory and Design*, IEEE Press Series on Electromagnetic Wave Theory, John Wiley & Sons, 2003.

[28] R. C. Hansen, "Wiley series in microwave and optical engineering," in *Phased Array Antennas*, 2nd ed., Wiley, 2009.

[29] R. J. Mailloux, *Phased Array Antenna Handbook*, 2nd ed., Artech House, 2005.

[30] R. L. Haupt, *Antenna Arrays: A Computational Approach*, Wiley-IEEE, 2010.

[31] P. Nayeri, A. Z. Elsherbeni, and F. Yang, "Radiation analysis approaches for reflectarray antennas," *IEEE Antennas and Propagation Magazine*, vol. 55, no. 1, pp. 127–134, 2013.

Index